T0145339

Springer Proceedings in Physics

Volume 231

Indexed by Scopus

The series Springer Proceedings in Physics, founded in 1984, is devoted to timely reports of state-of-the-art developments in physics and related sciences. Typically based on material presented at conferences, workshops and similar scientific meetings, volumes published in this series will constitute a comprehensive up-to-date source of reference on a field or subfield of relevance in contemporary physics. Proposals must include the following:

- name, place and date of the scientific meeting
- a link to the committees (local organization, international advisors etc.)
- scientific description of the meeting
- list of invited/plenary speakers
- an estimate of the planned proceedings book parameters (number of pages/ articles, requested number of bulk copies, submission deadline).

More information about this series at http://www.springer.com/series/361

Leonida Antonio Gizzi ·
Ralph Assmann · Petra Koester ·
Antonio Giulietti
Editors

Laser-Driven Sources of High Energy Particles and Radiation

Lecture Notes of the "Capri" Advanced Summer School

 Springer

Editors
Leonida Antonio Gizzi
Area della Ricerca del CNR
Istituto Nazionale di Ottica
Pisa, Italy

Petra Koester
Area della Ricerca del CNR
Istituto Nazionale di Ottica
Pisa, Italy

Ralph Assmann
Forschungszentrum
Helmholtz-Gemeinschaft
Deutsches Elektronen-Synchrotron DESY
Hamburg, Germany

Antonio Giulietti
Area della Ricerca del CNR
Istituto Nazionale di Ottica
Pisa, Italy

ISSN 0930-8989   ISSN 1867-4941   (electronic)
Springer Proceedings in Physics
ISBN 978-3-030-25852-8   ISBN 978-3-030-25850-4   (eBook)
https://doi.org/10.1007/978-3-030-25850-4

This Springer imprint is published by the registered company Springer Nature Switzerland AG
The registered company address is: Gewerbestrasse 11, 6330 Cham, Switzerland

In Memory

This volume is dedicated to Anatoly Faenov who left us in November 2017. Anatoly has been an unforgettable friend and an invaluable colleague during several decades of collaboration. His contribution to the field discussed in this volume has been truly outstanding and many seeds planted by his work continue to grow up and to produce new knowledge and new endeavours and projects.

We Editors will always remember Anatory's special attitude at collaborating with many groups worldwide, including us. We were awaiting each of his visits, always together with Tania, with great expectations and we always enjoyed his collaboration and his unique attitude to experimental work and exploration, always driven by curiosity. Often he used to come along with his magic X-ray spectrograph and surprise us with his ability of obtaining amazing space-resolved spectra of unprecedented resolution and contrast. Us, like many other groups around the globe, were proud to host his skill and work hardly day and night, enjoying each moment in the lab. More recently we were very enthusiastic to get with Anatoly and Tania electron and ion radiography using particles accelerated by laser interaction with clusters. He was always optimistic during experiments and actually he had good reasons for that; his deep knowledge and his terrific skill in making the experiment work, no matter how complex the setup was, made each measurement a success. His face, his smile, his dedication to the lab will remain a sweet memory forever.

Preface

New ultra-intense laser facilities with unique specifications, like the Extreme Light Infrastructure, will soon be on-line and will deliver laser performances never achieved before, based on the Chirped Pulse Amplification (CPA) concept [1]. It is worth mentioning that the extraordinary importance of CPA was recognised with the Nobel Prize in Physics in 2018, giving rise to a wave of interest from both the academic world and the general public for this novel science and technology. A number of dedicated laser installations are being built or upgraded across the world to enter new regimes of laser-matter interaction for particle acceleration and applications to generation of ionizing radiation. The major achievements in this field are now motivating the development of an entirely new generation of compact accelerator machines. This is a major leap that requires a new approach, similar to the approach followed by other pioneering enterprises like the development of radio-frequency accelerators occurred more than fifty years ago.

In this rapidly evolving context there is a compelling need of advanced training for the community of young researchers involved in the various aspects of this research, requiring theoretical, numerical and experimental skills.

The concept of the Advanced Summer School on *Laser-Driven Sources of High Energy Particles and Radiation* originates from the need of delivering advanced training in the field of novel acceleration techniques, gathering experts in optics, lasers, plasmas, accelerators and particle beams. The community of laser-plasma accelerator physics has traditionally been focusing on the extraordinary innovation emerging from ever-increasing laser intensities and performances and the advanced understanding of the physics of high-intensity laser-plasma interaction. Indeed, the School was also intended to provide tutorials on these fundamental physics aspects, introducing the theoretical framework and addressing the numerical approach to laser-plasma acceleration.

The School was conceived to bring together distinguished scientists and motivated young researchers, postdoc and Ph.D. students engaged or willing to enter this field, to promote advanced training in the key areas of ultraintense lasers, interaction with matter at ultra-high field and laser-plasma acceleration, with a focus on emerging, new, ground-breaking initiatives based on novel particle acceleration techniques, like the EuPRAXIA project aiming at the construction of the first ever industrial plasma- based particle accelerator. Novel accelerator techniques indeed were one of the highlights of the School, where the best expertise from the accelerator science was present to deliver fundamental notions and extraordinary achievements of modern light sources based on the latest acceleration technology, with a perspective view on novel accelerators. Also, specialists in generation of advanced radiation provided the latest update on applications to major multidisciplinary fields, including medicine and biology, material science and space industry.

Beam quality and reliability in this context call for a wide range of skills to be applied to overcome current issues in all aspects of laser-driven sources, from the laser driver, to the target plasma, developing appropriate diagnostic techniques for laser pulses, plasmas and beams and adopting appropriate stabilization and control techniques. All this goes through modelling of all components, from start to end, including all the relevant physical processes. This design approach, routinely followed by conventional accelerators, synchrotron and free-electron laser facilities, is now being extended to laser-plasma acceleration, defining a common knowledge base and promoting advanced training.

A special attention goes to the impressive developments of laser technology, now moving rapidly towards high average power systems, capable of higher repetition rate and gradually integrating efficient pumping technology. Progress has been very fast in this area, with laser labs promoting approaches ranging from evolution of existing technology, to entirely new schemes, based on laser materials capable of overcoming current limitations and aiming at performances compatible with the most advanced power drivers of RF accelerators. The School aims at delivering training in all these key areas of ultraintense lasers and laser-plasma interactions, having in mind the development of novel accelerator machines.

In summary, the scientific programme of the course covered all aspects of plasma acceleration, including fundamental laser-plasma interaction at high intensity, beam driven and laser-driven electron acceleration and different flavours of laser-driven ion acceleration. Leading experts guided participants through a journey across the science and technology of intense lasers, including advanced laser schemes for future high average power sources, fundamental aspects of laser-plasma interactions, electron beam dynamics and advanced configurations of radiation emission, from Thomson scattering to X-ray free electron laser. Ultrafast measurements and ultimate diagnostic techniques for laser and plasma characterization were a key part of the course, with step-by-step training on laser pulse amplification and compression, pulse duration and temporal contrast measurement, phase control and frequency conversion, electron and ion detection. Basic and advanced concepts of numerical modelling of laser-plasma interaction physics, radiation emission and particles and radiation transport were presented in view of the development of full start-to-end simulation of radiation sources. An overview was also given of the main laser facilities featuring the most advanced high power and high energy laser sources.

The School took place at the Conference Facility of the Consiglio Nazionale delle Ricerche (CNR) in Anacapri, on the island of Capri, in the south of Italy, off the coast of the beautiful peninsula of Sorrento. The conference facility is located at the premises of the former Solar Observatory of the Swedish Royal Academy, now owned by the Italian National Research Council. Capri is a famous destination in the Tyrrhenian Sea on the south side of the Gulf of Naples in the Campania region of Italy. Anacapri is located on the slopes of Mount Solaro at a higher elevation than Capri town. Points of interest include The Blue Grotto, the greatest attraction of the island, Villa San Michele, the Villas of Tiberius and the Chairlift for Mount Solaro which takes to the highest peak of the island from which you can enjoy the most stunning panoramas.

Pisa, Italy   Leonida Antonio Gizzi
Hamburg, Germany   Ralph Assmann
Pisa, Italy   Petra Koester
Pisa, Italy   Antonio Giulietti

Reference

1. D. Strickland, G. Mourou, Compression of amplified chirped optical pulses. Opt. Commun. **56**(3), 219–221 (1985)

Participants

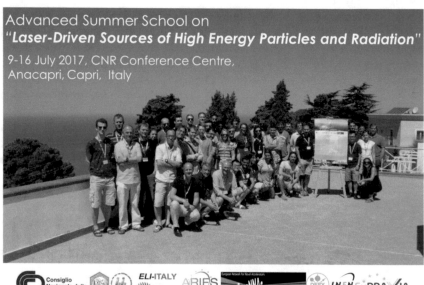

Fig. 1 Group Picture of the 2017 Capri Advanced Summer School

Roman Adam, Forschungszentrum Jülich, Germany
Carmen Altana, INFN, LNS, Italy
Christopher Baird, University of York, UK
Andy Bayramian, LLNL, Livermore, CA, USA (lecturer)
Carlo Benedetti, LBNL, Berkeley, CA, USA (lecturer)
Marco Borghesi, QUB, Belfast, UK, (lecturer)
Giada Cantono, LIDYL, CEA, CNRS, Universitè Paris-Saclay, France
Uddhab Chaulagain, ELI Beamlines, IoP, ASCR, The Czech Republic
Pablo Cirrone, INFN-LNS, Catania, Italy (lecturer)
Gemma Costa, INFN-LNF, Italy
Emma-Jane Ditter, Imperial College of S. T. & M., London, UK
Ángel Ferran Pousa, DESY, Hamburg, Germany
Massimo Ferrario, INFN-LNF, Italy (lecturer)
Rory Garland, Queen's University Belfast, UK
Elias Gerstmayr, Imperial College London, UK
Amin Ghaith, Paris-Sud University, Synchrotron SOLEIL, France
Dario Giove, INFN, Milano, IT (lecturer)
Jan-Niclas Gruse, Imperial College London, UK
Bernhard Hidding, SCAPA, University of Strathclyde, UK (lecturer)
Vojtěch Horný, Czech Technical University, Prague, Czech Republic

Malte Kaluza, IOQ, Jena, Germany (lecturer)
Masaki Kando, QST, JAEA, Kyoto, Japan (lecturer)
Stefan Karsch, MPQ, Munich, Germany (lecturer)
Kai Huang, Kansai Photon Science Institute, NIQRST, Japan
Luca Labate, CNR-INO, Pisa, Italy (lecturer)
Bruno Le Garrec, CNRS, Paris, France (lecturer)
Andrea Macchi, CNR-INO and Dip. Fisica, Pisa, Italy (lecturer)
Joel Magnusson, Chalmers University of Technology, Gothenburg, Sweden
Philip Martin, Queen's University Belfast, UK
Paul Mason, Central Laser Facility, STFC - RAL, Chilton, UK (lecturer)
Zeudi Mazzotta, CNRS, Laboratoire LULI, France
Aodhan McIlvenny, Queen's University Belfast, UK
Francesco Mira, Università degli studi di Roma "La Sapienza", Roma, Italy
Seyed Mirfayzi, Queen's University Belfast, UK
Annamaria Muoio, INFN, LNS, Italy
Zulfikar Najmudin, Imperial College of S. T. & M., London, UK (lecturer)
Ceferino Obcemea, NCI, Bethesda, MD, USA (lecturer)
Daniele Palla, CNR-INO, Pisa, Italy
Nicola Panzeri, Università degli studi di Milano, INFN, Milano, Italy
Gianfranco Paternò, INFN, Ferrara, Italy
Francesco Pisani, Università di Pisa, Dipartimento di Fisica, Italy
Savio Rozario, Imperial College London, UK
André Sobotta, Forschungszentrum Jülic Germany
Benjamin S. Wettervik, Chalmers U. of Technology, Gothenburg, Sweden
Elena Svystun, DESY, Hamburg, Germany
Davide Terzani, Università "Federico II", Naples, Italy
Riccardo Tommasini, LLNL, Livermore, CA,USA (lecturer)
Christopher Underwood, University of York, UK
Paul Andreas Walker, DESY, Hamburg, Germany
Longqing Yi, Chalmers U. of Technology, Gothenburg, Sweden

Contents

Contributors

Andrew Beaton Department of Physics, University of Strathclyde, Glasgow, UK; Cockcroft Institute, Sci-Tech Daresbury, Daresbury, Cheshire, UK

Marco Borghesi Centre for Plasma Physics, Queen's University Belfast, Belfast, UK

Lewis Boulton Department of Physics, University of Strathclyde, Glasgow, UK; Cockcroft Institute, Sci-Tech Daresbury, Daresbury, Cheshire, UK; Deutsches Elektronen-Synchrotron DESY, Hamburg, Germany

G. A. P. Cirrone INFN-LNS, Catania, Italy; ELI-Beamline Project, Institute of Physics, ASCR, PALS Center, Prague, Czech Republic

Sebastién Corde LOA, ENSTA ParisTech, CNRS, Ecole Polytechnique, Université Paris-Saclay, Palaiseau, France

G. Cuttone INFN-LNS, Catania, Italy

Alberto Martinez de la Ossa Deutsches Elektronen-Synchrotron DESY, Hamburg, Germany

Andreas Doepp Ludwig-Maximilians-Universität München, Garching, Germany; Helmholtz-Zentrum Dresden-Rossendorf, Institute of Radiation Physics, Dresden, Germany

M. Ferrario Frascati National Laboratory, National Institute for Nuclear Physics, Rome, Italy

Leonida Antonio Gizzi Istituto Nazionale di Ottica, Consiglio Nazionale delle Ricerche, Pisa, Italy; Sezione di Pisa, Istituto Nazionale di Fisica Nucleare, Pisa, Italy

Fahim Ahmad Habib Department of Physics, University of Strathclyde, Glasgow, UK; Cockcroft Institute, Sci-Tech Daresbury, Daresbury, Cheshire, UK

Thomas Heinemann Department of Physics, University of Strathclyde, Glasgow, UK;
Cockcroft Institute, Sci-Tech Daresbury, Daresbury, Cheshire, UK;
Deutsches Elektronen-Synchrotron DESY, Hamburg, Germany

Bernhard Hidding Department of Physics, University of Strathclyde, Glasgow, UK;
Cockcroft Institute, Sci-Tech Daresbury, Daresbury, Cheshire, UK

Arie Irman Max Planck Institut ü Quantenoptik, Garching, Germany

Malte C. Kaluza Institute of Optics and Quantum Electronics, Jena, Germany;
Helmholtz-Institute Jena, Jena, Germany

Masaki Kando Kansai Photon Science Institute, National Institutes for Quantum and Radiological Science and Technology, Kizugawa, Kyoto, Japan

Stefan Karsch Ludwig-Maximilians-Universität München, Garching, Germany;
Helmholtz-Zentrum Dresden-Rossendorf, Institute of Radiation Physics, Dresden, Germany

Gavin Kirwan Department of Physics, University of Strathclyde, Glasgow, UK;
Cockcroft Institute, Sci-Tech Daresbury, Daresbury, Cheshire, UK;
Deutsches Elektronen-Synchrotron DESY, Hamburg, Germany

Alexander Knetsch Deutsches Elektronen-Synchrotron DESY, Hamburg, Germany

Luca Labate Istituto Nazionale di Otticam, Consiglio Nazionale delle Ricerche, Pisa, Italy;
Sezione di Pisa, Istituto Nazionale di Fisica Nucleare, Pisa, Italy

Bruno LeGarrec LULI, Ecole Polytechnique, Palaiseau, France

Andrea Macchi National Institute of Optics, National Research Council (CNR/INO), Adriano Gozzini laboratory, Pisa, Italy;
Enrico Fermi Department of Physics, University of Pisa, Pisa, Italy

Grace Gloria Manahan Department of Physics, University of Strathclyde, Glasgow, UK;
Cockcroft Institute, Sci-Tech Daresbury, Daresbury, Cheshire, UK

D. Margarone ELI-Beamline Project, Institute of Physics, ASCR, PALS Center, Prague, Czech Republic

Zulfikar Najmudin Blackett Laboratory, Department of Physics, The John Adams Institute for Accelerator Science, Imperial College London, London, UK

Alastair Nutter Department of Physics, University of Strathclyde, Glasgow, UK;
Cockcroft Institute, Sci-Tech Daresbury, Daresbury, Cheshire, UK;
Max Planck Institut ü Quantenoptik, Garching, Germany

L. Pandola INFN-LNS, Catania, Italy

G. Petringa INFN-LNS, Catania, Italy

Paul Scherkl Department of Physics, University of Strathclyde, Glasgow, UK; Cockcroft Institute, Sci-Tech Daresbury, Daresbury, Cheshire, UK

Ulrich Schramm Max Planck Institut ü Quantenoptik, Garching, Germany

Daniel Ullmann Department of Physics, University of Strathclyde, Glasgow, UK; Cockcroft Institute, Sci-Tech Daresbury, Daresbury, Cheshire, UK; Central Laser Facility, STFC Rutherford Appleton Laboratory, Didcot, Oxfordshire, UK

Chapter 1
Laser-Driven Sources of High Energy Particles and Radiation

Leonida Antonio Gizzi

Abstract Ultraintense lasers are now established as powerful drivers for high energy particle, plasma based accelerators and unique compact radiation sources. Further developments will require the coalescence of a wide range of fields, from optics and lasers, to plasmas and particle beams, to deliver the first accelerator with reliable and reproducible operation and with extraordinary specifications. This volume collects papers from key experts, covering the most crucial topics and offering young researchers the reference source required to identify fundamental physical aspects and the most advanced technology. In this chapter I will highlight some of the most exciting developments of the field and their perspective, also in view of multi-disciplinary applications.

1.1 Introduction

Since the "dream beam" results [1–3] emerged after the first impressive exploitation of ultraintense lasers, laser-plasma acceleration has been developing at an unprecedented pace, with the most recent achievement of record electron energy exceeding 8 GeV [4]. Equally important is the demonstration of staging [5], were the concept of acceleration module, required for scaling to high energy, is first put forward. At the same time, effort continues in the understanding of the electron injection process, separated from the wakefield generation and the subsequent acceleration process. In fact, the major breakthrough in electron beam quality is expected from the control of the electron injection to minimize beam emittance and energy spread, two of the most important quality factors of the accelerated bunch. Here several milestones have been achieved using different physical processes, and injection schemes continue to evolve (see [6] and references therein), moving towards an "optical engineering"

L. A. Gizzi (✉)
Istituto Nazionale di Ottica, Consiglio Nazionale delle Ricerche, Pisa, Italy
e-mail: la.gizzi@ino.cnr.it
URL: http://www.ilil.ino.it

Sezione di Pisa, Istituto Nazionale di Fisica Nucleare, Pisa, Italy

© Springer Nature Switzerland AG 2019
L. A. Gizzi et al. (eds.), *Laser-Driven Sources of High Energy Particles and Radiation*,
Springer Proceedings in Physics 231, https://doi.org/10.1007/978-3-030-25850-4_1

approach of the process where laser and plasma parameters are fine-tuned using accurate three-dimensional numerical simulations. In view of this, the control over laser and plasma input parameters and their stability is of paramount importance.

In parallel, a major effort is being dedicated to the electron beam transport and secondary radiation generation via several mechanisms, ranging from simple target impact for Bremmstrahlung emission [7] and positron generation [8], to Thomson scattering for γ-ray emission [9], to undulator for X-ray free electron laser generation. The latter requires a demanding combination of electron bunch specifications and is currently among the main scientific objectives of the major laser-plasma acceleration labs. From the point of view of transport of accelerated electrons, plasma lenses are being studied [10] as powerful, tunable devices capable of cm-scale focal lengths for high energy beams, thus allowing compactness of laser-plasma accelerators to be preserved, while effectively transporting the electrons from one stage to another.

In this Chapter a review will be given of the main areas relevant for the development of "Laser-Driven Sources of High Energy Particles and Radiation", highlighting the current trend and identifying the science and technology paths currently being followed to achieve the next milestones in Laser-Plasma Acceleration, Plasma Diagnostics, Radiation Sources, Laser Drivers for Plasma Accelerators and Multidisciplinary Applications. The reader is addressed to the following chapters for a detailed tutorial description of the main topics.

1.2 Laser-Plasma Acceleration

This was one of the main topics of the School and the core of this book includes four lectures addressing the fundamental aspects of high intensity laser-plasma interactions (see Lecture by A. Macchi) and acceleration (see Lecture by Z. Majmudin), the extension of the concept to staged acceleration (see Lecture by M. Kando) and the outstanding hybrid concept of particle wakefield acceleration driven by a laser-driven electron beam (see Lecture by B. Hidding). Given the weight of this topic in the School, a short introduction is given here to recall the main advantages and the current limitations of laser-plasma acceleration, while bearing in mind the necessary comparison with high brightness electron beams and their current status in conventional accelerators, as described in another tutorial lecture of the School (see lecture by M. Ferrario).

In the classical picture of Laser Wakefield Acceleration (LWFA) [11], a longitudinal electron plasma wave is excited by the ponderomotive force associated to an ultra-short, ultraintense laser pulse. The electron plasma wave is characterized by a longitudinal electric field and a phase velocity set by the group velocity of the laser pulse, $v_g = c(1 - \omega_p^2/\omega_L^2)^{1/2}$, where $\omega_p = (n_e e^2/\epsilon_o m_e)^{1/2}$ is the electron plasma frequency, with n_e being the electron plasma density, e, m_e and ϵ_o the electron charge and mass and the dielectric constant respectively and ω_L is the laser angular frequency. Electrons in phase with the wave are accelerated until, travelling faster than the electron plasma wave, overcome the accelerating field of the wave and start experiencing a decelerating field. This mechanism yields a maximum accelerating distance equal to the so-called dephasing length, given by:

$$L_d = \frac{\omega_L^2}{\omega_p^2}\lambda_p \simeq 3.2 \; n_{18}^{-3/2} \; \lambda_{\mu m}^{-2}, \tag{1.1}$$

where n_{18} is the electron density in units of 10^{18} cm^{-3} and $\lambda_{\mu m}$ is the laser wavelength in μm. At high laser intensity, this classical scenario is significantly modified and numerical simulations provide detailed description of plasma wave excitation and evolution as well as electron injection and acceleration. The most compact configuration to obtain GeV-range electron bunches from laser-plasma interaction is based upon a gas-jet of a few millimeters, working in the so-called blowout regime [12, 13]. A short ($c\tau < \lambda_p/2$) and intense ($a_0 > 2$) laser pulse expels the plasma electrons outward creating a bare ion column. The blown-out electrons form a narrow sheath outside the bubble and the space charge generated by the charge separation pulls the electrons back creating a bubble-like wake whose size is $\lambda_b \simeq 2\sqrt{a_0}c/\omega_p$ and the de-phasing length becomes $L_d = 2/3(\omega_L^2/\omega_p^2)\lambda_b$. Here $a_0 = eA/mc^2 = 8.5 \times 10^{-10}\sqrt{I\lambda^2}$ is the normalized vector potential of the laser, I and λ being the laser intensity in W/cm^2 and the laser wavelength in μm. For sufficiently high laser intensities ($a_0 > 3.5 \div 4$) electrons at the back of the bubble can be injected in the cavity and experience a maximum accelerating field of $E_{acc}[GV/m] \simeq 100 \; n_{18}^{1/2}$. It can be shown that the maximum energy gain is given by [14]

$$W_{max}[GeV] = E_{acc}L_d \simeq 0.37 \; P_{TW}^{1/3} \; n_{18}^{-2/3}, \tag{1.2}$$

According to this result, a matched condition (acceleration over the entire dephasing length) to achieve 1 GeV electron energy, at a moderate laser power of 80 TW, requires an electron density of 2×10^{18} cm^{-3}. At this relatively high electron density, experiments show that laser beam quality is the key parameter to enable a satisfactory propagation, but a range of processes still play a crucial role in the propagation. Diagnostic techniques aimed at characterizing the propagation dynamics and unveiling the microscopic features of accelerating structures in the plasma are therefore needed to gain control over the acceleration process. In this context, special attention is being dedicated to the control of self-injection of electrons. Recently, several mechanisms have been identified and implemented to control injection of electrons in a well-formed wake wave which can be broadly divided into three categories depending on the basic physical mechanism responsible for injection. The objective is to achieve a localized injection of electrons with a limited longitudinal spatial extent, to ensure reduced energy spread of accelerated electrons. Wave breaking is certainly the most fundamental process leading to injection of electrons in a plasma wave. While transverse wave breaking [15] suffers from a de-localized injection of electrons and consequently large energy spread, longitudinal wave breaking via down-ramp [16] density-transition [17] certainly provides more localized injection and limited energy spread of electrons. Activation of such injection schemes required accurate control of shape and profile of electron distribution that can be achieved using custom gas targets and plasma tailoring. Recent successful implementations of this principle yielding very localized injection have been demonstrated which rely on plasma lensing [18] and shock-front in gas jets [19]. Ponderomotive injection [20, 21] also enables a high degree of control on the exact location of injection, but requires significantly more complex experimental configurations with additional laser pulses.

A conceptually simple technique to enhance electron injection is the so-called ionization-injection [22] in which field ionization properties of some gases are exploited to increase electron density in the bubble. Recent advances of this scheme also enable control of the spatial distribution of ionization injection and consequent smaller energy spread.

Indeed, it is the dramatic development of these injection techniques which is currently enabling generation of narrow energy spread electrons with high energy, up to the multi GeV uniquely by laser techniques.

In the two-color ionization injection [23] two laser pulses are used. The main pulse has a long wavelength, five or ten micrometers, and a large normalized amplitude, $a_0 > 1$ and drives the plasma wave in a low Z gas with a medium Z dopant. The second pulse, the ionization pulse, is a shorter wavelength, typically a frequency doubled, 400 nm Ti:Sa pulse. While the main pulse cannot further ionize the electrons in the external shells of the large Z dopant due to its large wavelength, the electric field of the ionization pulse is large enough to generate newborn electrons that will be trapped in the bucket. This opens the possibility of using gas species with relatively low ionization potentials, thus enabling separation of wake excitation from particle extraction and trapping. The main drawbacks of the two-color ionization injection are the lack of availability of short, intense 100 TW-class laser systems operating at large ($\approx 10\,\mu$m) wavelength and lasers synchronization jitter issues. These limitation make the two-color scheme currently unpractical for application to LWFA-based devices that aim at high quality beams. More recently, a novel scheme was proposed, the Resonant Multi-Pulse Ionization injection [6] in which instead of the long wavelength driving pulse of the two-color scheme is replaced by a short wavelength, resonant multi-pulse laser driver. Due to the resonant enhancement of the ponderomotive force, a properly tuned train of pulses is capable of driving amplitude waves larger than a single pulse with the same total laser pulse energy. Noticeably, since the peak intensity of the driver is reduced by a factor equal to the number of train pulses, it is also possible to match the conditions of both particle trapping and unsaturated ionization (i.e. with low ionization percentage) of the active atoms level. In this way, the practical limitations of the two-color ionization injection can be overcome. Based on this scheme, accurate numerical simulations show that a GeV accelerator with a normalized emittance of 80 nm × rad and an rms energy spread below 0.5% could be achieved, as shown in Fig. 1.1 [24]. More effort is needed in this direction and perspectives in the near future are that injection and acceleration up to the 5 GeV energy range will be stable and accurate as required to drive a new generation of radiation sources for applications.

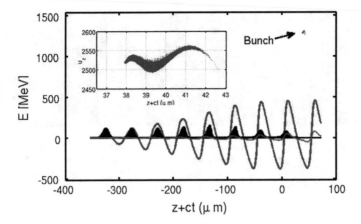

Fig. 1.1 Longitudinal phase space of an electron bunch (red dot) accelerated with the Resonant Multi-Pulse Ionization Injection [6], after 3.7 cm of propagation is the plasma. The blue line shows the electric field on axis (a.u.), while the red line represents the transverse focusing force at a radius close to the beam radius (a.u.) [24]

1.3 Plasma Diagnostics

Laser-plasma acceleration relies heavily on the interaction of laser pulses with gases and plasmas and, currently, the focus is on the reproducibility of the propagation process to minimize effects on the acceleration process and the quality of the accelerated electron bunches. Indeed, a lot remains to be investigated in this context concerning the study of temporal and spatial features of the plasma during propagation, and the school delivered tutorials on this crucial aspect (see Lecture by M. Kaluza).

As an appetizer, a basic discussion of plasma probing is given here on techniques used to determine basic plasma parameters and to detect interaction processes during propagation. These techniques are generally referred to as optical plasma probing techniques and include schemes known as shadowgraphy, knife-edge (Schlieren), interferometry and polarimetry (e.g. Faraday rotation). These schemes can provide a map of basic plasma parameters and, if a pulsed probe beam is used with adjustable time delay, temporal evolution can also be achieved. A full description of optical probing techniques can be found elsewhere [25]. Here we provide a basic introduction with specific relevance to laser-plasma acceleration.

Optical probing techniques have been used in the past decades to diagnose plasmas produced by nanosecond laser pulses in laser-produced plasmas and in laser-fusion related experiments as discussed in [26]. The image of Fig. 1.2 shows a pioneering example of shadowgraphy of laser produced shock waves in gas and the use of holographic interferometry to study early stages of spark creation by laser, converging shock waves and high Mach number shocks induced by collision of two spherical shocks. The conceptual set up for a plasma probing experiment is shown in Fig. 1.3 with the "main" laser pulse incident on a plasma and the pulse propagating through

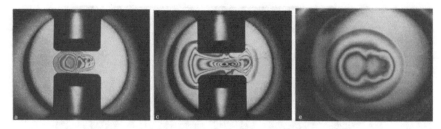

Fig. 1.2 An example of a double-pulse holographic interferometry of plasma generated by a GW laser pulse focused in a gas, showing the initial expansion (left) and convergence of a shock after reflection off a flat surface (center). The interferogram of the freely expanding plasma is also shown for comparison (right). Such interferometric techniques were crucial in this class of experiments to map the electron density and its temporal evolution [27]

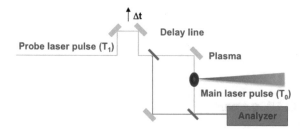

Fig. 1.3 Conceptual set-up of optical probing to study laser propagation in a plasma. The relative timing between the arrival on the target plasma of the main laser pulse (T_0) and the probe pulse (T_1) can be controlled using a time-slide (Δt). The probe pulse is then analyzed against the reference pulse to extract phase changes due to propagation in the plasma due to refraction, phase shift, polarization etc.

the same plasma along a preferred direction (e.g. perpendicular to the propagation axis of the main laser pulse). The probe pulse undergoes changes of its properties due to interactions with the plasma. Since the intensity of the probe is typically very small (well below the intensity at which non-linear effects take place), the probe pulse will mainly suffer changes due to the refractive properties of the plasma. In order to detect density gradients in the plasma, the so-called Schlieren technique is used in which a collimated probe pulse propagates through the plasma and is then focused using a lens. In the focal region, a sharp (knife) edge is used to block all the unperturbed probe rays going through the focal point. The remaining probe rays, deflected by their original path by deflection in the plasma, will then be used to form an image of the plasma in which all density gradients are visible. This approach provides a qualitative analysis of the plasma density map, showing density perturbations in a very effective way.

A quantitative analysis of the plasma density map can be obtained using plasma interferometry. In this case, the probe pulse that has propagated through the plasma will be analyzed to measure the phase shift induced by the plasma refractive effects. The plasma refractive index of the electron density given by

$$\eta = \sqrt{1 - n_e/n_c} \tag{1.3}$$

with n_e and n_c being the electron density and the critical density respectively. For $n_e << n_c$ the above relationship becomes

$$\eta \simeq 1 - n_e/2n_c. \tag{1.4}$$

According to (1.4) above, a ray of an optical beam passing through the plasma will acquire a phase shift with respect to a ray which has travelled the same distance in a vacuum given by:

$$\Delta\phi = -(2\pi/\lambda) \int_L [\eta(x, y, z) - 1] \, dx \simeq (\pi/\lambda n_c) \int_L n_e dx. \tag{1.5}$$

By measuring the phase shift, information about the electron density can therefore be obtained. The final expression of the phase shift [26] belongs to a family of integrals called Abel Integrals and for cylindrical symmetry can be inverted analytically to extract the electron density. The resulting phase shift map will therefore be retrieved using Abel inversion under a suitable assumption on plasma symmetry to generate the density map. Interferometric techniques rely on the assumption that the electron density is well below the critical density and that transverse density gradients are small enough so that the probe pulse can propagate through the whole plasma without undergoing severe deflection.

Interferometric arrangements consisting of a Mach-Zehnder or a Nomarsky type interferometer are shown schematically in Fig. 1.4, producing a fringe pattern. In the Nomarski interferometer, a probe beam is set to propagate through the region of interest. A lens is used to image the region of interest with the required magnification and resolution. A Wollaston prism is then used to split the probe beam into two partially overlapping beams, having relative orthogonal polarization and that appear to be emerging from two separate virtual foci. As a consequence two laser beam spots are projected on the detector, each enclosing an image of the plasma. By

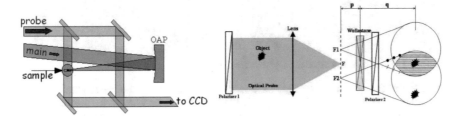

Fig. 1.4 (left) A schematic layout of the Mach-Zehnder (left) and Nomarski (right) interferometer analysis of plasmas. While the first is based on the classical two-arm configuration as shown in Fig. 1.4 and provides a full aperture interferometric image of the plasma, the second provides a simplified set-up due to its simple apparent "single" arm configuration

appropriately setting the distance between the prism and the detector, the plasma image from one of the spots is made to overlap with an unperturbed region of the other spot. In the overlapping region interference is produced with the use of a pair of polarizer selecting the same polarization and nearly equal intensities. In the typical arrangement, P1 selects a linearly polarized component of the incoming probe beam, while P2 is mounted with the polarization axis rotated by 45 with respect to the axis of P1. In the overlapping region an interference pattern is produced with a fringe separation given by: $\Delta z = \lambda p/\alpha q$, where λ is the wavelength of the optical probe, α is the angular aperture of the Wollaston prism, and q and p are the distances between the prism and the detector plane, and between the Fourier plane of the lens and the prism respectively. The fringe separation can be modified by changing the position of the Wollaston prism relative to the lens. Figure 1.5 shows the interferogram of a pulsed gas-jet in vacuum acquired by means of a Nomarski interferometer. As discussed in [26] numerical techniques can be used for the fringe pattern analysis to extract phase-shift information as small as a fraction of wavelength and retrieve the density map shown in the lower panel. The interferogram shows the positive phase shift due to the refractive index of neutral atoms arising from the interaction of the probe pulse with bound electrons and resulting in a small refractive effect. In contrast, in a plasma, the contribution of the free electrons to the phase shift is dominant. The image of Fig. 1.6 shows the effect of the plasma on the probe pulse, visualizing the propagation of a femtosecond, high intensity laser pulse from right to left, across the focal region. Interestingly, further refining of these techniques in the context of laser plasma acceleration has brought to the visualization of the wakefield formation and generation in the plasma, as discussed in details in Chap. 7 of this book.

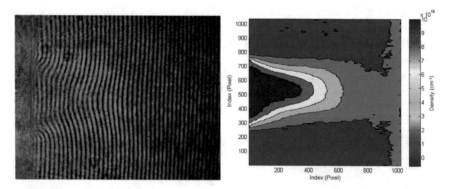

Fig. 1.5 Interferogram of a pulsed gas-jet in vacuum with a rectangular nozzle, acquired by means of a Nomarski interferometer [25] (left). The red marker shows the width of the nozzle. The corresponding average density map obtained with a phase-retrieval algorithm [28] is also displayed (right)

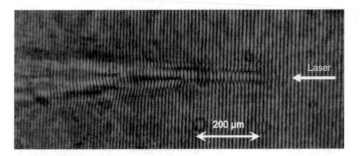

Fig. 1.6 Interferogram of the laser-plasma interaction region in a gas-jet interaction experiment. The shot was taken in N_2 at a backing pressure of 35 bar, with the laser focal plane located 50 μm into the gas-jet and 400 μm from the nozzle output plane. The interferogram was taken 2 ps after propagation, from right to left, of the laser pulse, showing the laser channeling through the focal region

1.4 Radiation Sources

A new perspective for compact, all-laser driven X-ray and γ-ray sources is emerging, aiming at a brightness currently achievable only with large scale facilities like synchrotron radiation facilities. Bremsstrahlung or fluorescence emission driven from fast electron generation in laser interaction with solids was demonstrated to provide effective ultrashort X-ray emission with unique properties. On the other hand, laser-plasma electron acceleration is being considered in place of conventional radio-frequency electron accelerators for a variety of radiation emission mechanisms. Broadband radiation generation schemes including betatron and Bremsstrahlung are being developed while Thomson scattering by collision with a synchronized laser pulse is being proposed for the generation of narrow band radiation. Moreover, with the increasing quality of laser-driven electron bunches and the expected availability of high repetition rate laser drivers, X-ray free electron lasers are also being designed [29].

Bremsstrahlung emission is the simplest mechanism, based on Coulomb collisions of charged particle with the nucleus of an ion, undergoing acceleration and emitting radiation with a continuous photon energy spectrum that extends up to approximately the electron rest energy times the γ factor of the incident electron. For values of $\gamma > 1$, photons are emitted in the forward direction in a cone of aperture of approximately $1/\gamma$. The total radiated power scales as Z^2 and can account for a conversion of a significant fraction of the electron energy into photon energy. Practical Bremsstrahlung sources extend from the keV range, as in the X-ray tube, up to the multi MeV range. In the latter case, the high energy electron bunch, accelerated by a linac, hits a converter, typically a tungsten or a tantalum plate, and generates γ-rays with photon density as high as 1 ph/eV/sec. Alternatively, high energy electron bunches produced using compact plasma accelerators driven by lasers can be used in place of linac generated electrons. All-optical, laser-based bremsstrahlung

X-ray and γ-ray sources have already been explored [30, 31] and successfully tested using self-injection electron bunches [7, 32]. A typical laser-driven γ-ray source was used in [33] to activate a gold sample in the 8–17.5 MeV photon energy range of the giant dipole resonance. A total flux of 4×10^8 photons per Joule of laser energy was estimated through activation measurements which makes this class of sources the brightest Bremsstrahlung source in the considered photon energy range [34]. Photon yield can be significantly enhanced if multiple bunches are generated in laser wakefield acceleration for a single laser pulse. In fact, in the experimental conditions explored in [34], the laser pulse undergoes self-phase modulation and compression that leads to the excitation of a non-linear plasma wave with multiple buckets with a similar amplitude. Injection and acceleration occurs in each bucket and consequently, multiple electron bunches of high energy electrons are generated at each laser shot. When a similar set up is considered for applications like imaging and non destructive testing of thick objects, the source size is a very relevant parameter that must be optimized to enhance spatial resolution of the imaging technique. Recent studies [35] show that a source size as small as 30 μm can be obtained placing the Bremsstrahlung converter a few millimetres from the gas-jet downstream the bunch propagation direction to perform high resolution γ-ray imaging of bulky and dense objets [36]. An additional feature of all-laser driven sources is the intrinsic ultrashort pulse duration which, combined with the potentially high degree of compactness, makes this class of sources unique and potentially advantageous for applications in a wide range of fields, both in industry and in basic research.

Betatron emission is another very effective mechanism of generation of X-ray radiation during laser-plasma wakefield acceleration, originating from the transverse oscillation of the electron bunch in the acceleration cavity due to the strong transverse restoring force. The typical photon energy of this emission is similar to the undulator type radiation characterised by a wavelength of the oscillation $\lambda_\beta = \sqrt{2\gamma_0}\,\lambda_p$, where $\lambda_p = 2\pi c/\omega_p = 1.05 \times 10^{21} n_e^{-1}$ μm is the electron plasma wavelength, with n_e being the electron plasma density in units of cm^{-3}. Photon energy up to the keV range can be easily achieved for electron energy up to 1 GeV. This radiation mechanism was first observed in 2004 [37] and is characterized by a small source size, enabling phase contrast imaging, as recently demonstrated both at 5 keV [38] and earlier at higher photon energies of 10 keV [39] and above 20 keV approximately [40]. Although betatron emission typically exhibits a broadband spectrum in the keV range, higher photon energy can be achieved in conditions of enhanced transverse electron oscillations as demonstrated in [41] where parameters of the accelerating cavity were modified in such a way to enable overlapping of the electrons with the rear of the laser pulse. In this way the betatron motion was resonantly excited and the resulting oscillation amplitude was found to increase significantly, leading to an enhanced X-ray photon energy.

Thomson scattering is being considered for the generation of higher energy photons when a high intensity laser pulse is set to collide with a relativistic electron bunch. It was initially proposed in 1963 [42, 43] as a quasi monochromatic and polarized photons source. With the development of ultra intense lasers the interest on this process has grown and the process is now being exploited as a bright source

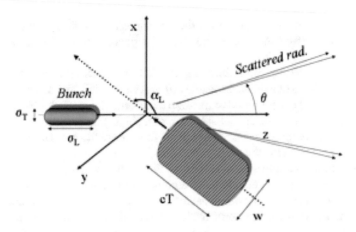

Fig. 1.7 Thomson scattering geometry. The scattered radiation is emitted along the z axis, in a small cone of aperture $1/\gamma$. When $\alpha_L = \pi$ the backscattering geometry occurs

of energetic photons from UV to γ-rays and atto-second sources in the full nonlinear regime.

We consider the geometry described in Fig. 1.7. The three main parameters governing the scattering process are the electron energy $E_o = \gamma_o m_e c^2$, the laser pulse peak normalized amplitude $a_o = eA/(m_e c^2) \approx 8.5 \times 10^{-10} \sqrt{I \lambda_{\mu m}^2}$, I being the laser peak intensity in W/cm^2, $\lambda_{\mu m}$ is the laser wavelength in μm and α_L is the angle between the propagation directions of the laser pulse and the electrons.

The pulse amplitude controls the momentum transferred from the laser pulse to the electron, i.e. the number of photons of the pulse absorbed by the electron. If $a_o << 1$, only one photon is absorbed and the resulting electron motion always admits a reference frame in which the quivering is non-relativistic (linear Thomson scattering) [44]. Assuming $\gamma_o >> 1$, scattered radiation is emitted forward with respect to the electron initial motion within a cone of aperture $1/\gamma_o$. Assuming a laser pulse having a rise time much greater than the pulse period, the resulting scattered radiation ω_γ is spectrally shifted compared to the laser frequency ω_L at a peak energy given by [45]:

$$\omega_\gamma \cong 2\gamma_o^2(1 - \cos\alpha_L)\omega_L \tag{1.6}$$

Among the possible interaction geometries, the case of backscattering $\alpha_L = \pi$ is the most suitable for at least three aspects: (i) it produces photons with the highest energy

$$\omega_\gamma \cong 4\gamma_o^2\omega_L; \tag{1.7}$$

(ii) it allows the highest overlap of the electron beam and the pulse; (iii) it minimizes spurious effects induced by the transverse ponderomotive forces of the laser pulse.

Thomson scattering in the linear regime has also been proposed to attain the angular distribution of a monochromatic electron bunch [46]. Moreover, experimental methods have been proposed to measure the length of a monochromatic electron bunch and to measure the energy spectrum of a single bunch eventually characterized by a wide energy spread or alternatively to measure the angular distribution of a single bunch with a known energy spectrum [47].

In the nonlinear regime, $a_0 \approx 1$, the resulting strong exchange between the laser pulse and electron momentum induces a complex and relativistic electron motion, consisting of a drift and a quivering having both longitudinal and transverse components with respect to the pulse propagation. In turn, the time dependent longitudinal drifting results in a non-harmonic electron motion that produces scattered radiation with a complex spectral distribution characterised by harmonics of the fundamental frequency. If the electron interacts with a laser pulse with a constant amplitude, e.g. a flat-top laser pulse, the spectral distribution of the scattered radiation consists of equally spaced harmonics [44]. In the case of head-on collision, the peak energy of each Nth harmonics in a back-scattering configuration reads now:

$$\omega_{\gamma,N} \cong N_{th} \frac{4\gamma_o^2 \omega_L}{1 + a_o^2/2}. \tag{1.8}$$

As the intensity increases even further ($a_0 \gg 1$), radiation is emitted into many closely spaced harmonics showing a typical synchrotron radiation spectrum. When considering scattering from an electron bunch, harmonics produced by each electron will be slightly shifted due to non-ideal beam effects like energy spread and beam emittance. As a consequence, a continuous spectrum is generated which extends up to the critical frequency that scales [48] as a_0^3.

Thomson scattering with laser driven electrons was first demonstrated in a pioneering experiment [49], showing evidence of 1 keV X-ray emission. Current experiments typically use a collision point set a few mm downstream the accelerating region, where electron bunch transverse size is a few tens of μms, easily achievable with the scattering pulse. In the original paper by Chen et al., [50], the collision point was set 1 mm after the exit of the plasma, where the focal spot of the 800 nm scattering pulse spot size was 9 μm and the overlapping (emitting) region was estimated to be 5 μ. Scattering with a 250 MeV cut-off energy electrons enabled generation of peak photon energy of 1.2 MeV. Sarri et al., [34] used a F/2 off-axis parabolic mirror to focus the 18 J, 42 fs scattering pulse 10 mm downstream of the exit of the gas target were the electron bunch transverse size was 30 μm and the average normalized intensity was $a_o = 2$. Scattering off LWFA electrons with energy up to 600 MeV resulted in Thomson scattering photons with energy up to 18 MeV, the highest energy obtained so far with all-optical Thomson scattering. Liu et al. [51] achieved similar photon energy with lower peak electron energy, but using frequency doubled, 400 nm optical scattering pulse. They used a separate optical compressor to control focal spot quality of the frequency doubled pulse which is more sensitive to phase front distortions. They tuned the frequency doubled scattering pulse to produce 54 mJ in a 300 fs laser pulse, focused in a 15 μ focal spot and were able to achieve >9 MeV photon energy with a broadband spectrum peaked at approximately 400 MeV.

Depending on the perspective application, tuneability of the X-ray photon energy may be an important option of a source. According to (1.7) or (1.8), the frequency of the scattered radiation can be tuned by changing either the electron energy or the scattering photon energy. Powers et al., [52] achieved tuneability changing the electron energy in the range from 50 to 200 MeV by changing the plasma density to exploit the square root dependence of the accelerating electric field upon the electron density that occurs in LWFA. In this way they were able to achieve tuneability in the range from 70 keV to approximately 1 MeV. Tunability in the 5–42 keV was demonstrated by Khrennikov et al., [53] using a different technique to tune electron energy. They use shock-front injection [19]which exploits the properties of sharp downramps of the electron density of the plasma [47] to localize electron injection. Tunability is achieved by shifting the position of the down-ramp along the plasma to control acceleration length.

1.5 Laser "Drivers" for Plasma Accelerators

High power laser technology has been developing rapidly after the first laboratory demonstration of laser operation [54], leading to an increase of peak power up to the multi-PW level and focused laser intensities of approximately 15 orders of magnitude, as shown in the plot of Fig. 1.8. Commercial laser systems are now available to reach such performances and many laboratories world-wide feature PW-scale laser systems [55] that are used as drivers of laser-plasma acceleration experiments. Further progress is in order, following progress in amplification architectures and diagnostic techniques. Both these aspects have been treated in details during the school and have dedicated chapters of this book (see lectures by B. Le Garrec and L. Labate). In the Section below a brief description of the leading laser technologies is given in the perspective of designing a driver for a laser-plasma accelerator.

The vast majority of such systems are based on Titanium Sapphire technology, where titanium doped Al_2O_3 (Ti:Sa)'s broad gain bandwidth allows pulses as short as a few tens of fs to be amplified at a wavelength of about 800 nm, making these systems perfectly suited for LWFA drivers. On the other hand Ti:Sapphire must be pumped in the visible (typically in the range 500–550 nm), which is commonly obtained by using frequency doubled, flashlamp-pumped Q-switched lasers operating in the near infrared (e.g. Nd:YAG). These circumstances strongly limit the wall-plug efficiency of the whole system making the high average power operation very challenging and currently limiting the repetition rate typically to $1/P_{PW}$ [Hz], where P_{PW} is the peak power in PW. The plot of Fig. 1.9 shows the laser specifications of some of the main facilities plotted according to their repetition rate and their energy per pulse. In this

Fig. 1.8 History of
achievable focused laser
intensity since the invention
of the laser. Main steps to
generate high laser power
include the so called
Q-switch technique and the
mode-locking and, more
recently the chirped pulse
amplification technique [56]

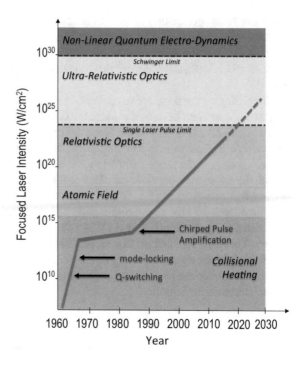

chart a set of fixed average power levels are represented by the parallel lines. It is
currently established that the average power required for the main applications of
high intensity lasers is above the kW range, namely, well above the values achieved
today, implying that a significant step in technology is needed to fill the gap of two
orders of magnitude in the deliverable average laser power.

Laser-based plasma accelerators would ideally require a typical repetition rate
of 1 kHz that can only be achieved provided diode-pumped solid state (DPSSL)
lasers are used in place of flash-lamp pumped technology. A significant improvement
in wall-plug efficiency can be obtained for Ti:Sa technology, replacing flash-lamp
pumped Nd lasers with DPSSL technology, like the recently established HAPLS
laser system [57] that aims at operation at an average power of 300 W at 10 Hz.
In the mean time, DPSSL pump lasers capable of delivering kW average power at
the required wavelength of 0.5 μm for petawatt scale systems are currently being
developed and full design of such systems is in progress for 100 Hz repetition rate
[58].

A crucial role in the development of high average power lasers is played by the
cooling architecture and geometrical layout of the multi-pass amplifier required to
manage heating of the gain material due to the pump energy. Studies are emerging
which show that amplifying media need to be water-cooled with high flow speed
in order to remove the heat generated by the pumping process. For this purpose,
amplifying media must be shaped as thin disks, with high diameter/thickness ratio
and cooling must be applied on the crystal faces in order to have a sufficient heat

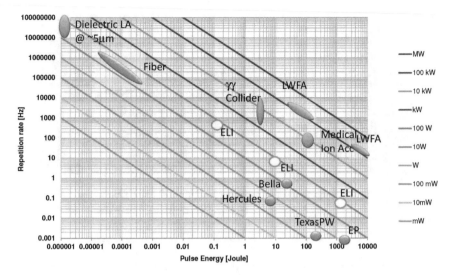

Fig. 1.9 Chart showing laser specifications of main facilities currently existing or under construction, plotted according to their repetition rate and to the energy per pulse. The parallel lines in the plot indicate a constant average power. Main applications require average power typically above the kW level, corresponding to repetition rate higher than 100 Hz and energy per pulse higher than 100 J [59]

exchange surface. To achieve sufficient pump absorption and energy storage, the required gain volume must be split in several disks depending on the pump energy and repetition rate, and to achieve efficient amplification and energy extraction the amplified beam must cross the volume several times (multi-pass). As for the cooling strategy, two possible approaches are considered, namely a transmission geometry and a reflection geometry. In the transmission geometry, the amplifying crystal is water-cooled on both faces, and both the pump beams and the amplified beam cross the crystal, the water-cooling flow, and the flow containment windows. It offers a good performance in terms of heat extraction and allows to implement simpler layouts from the geometrical point of view, but it presents a potential drawback because the amplified beam crosses the cooling flow and is potentially subjected to optical aberrations due to turbulences, as the fluid flows at high values of the Reynolds number. In the reflection geometry, one of the faces of the crystal is highly reflective for the amplification beam. The amplified beam enters in the crystal from the front face and it is reflected back in the incoming direction on the back surface. The same occurs to the pump beams. The reflective surface is water-cooled, whereas the front face is uncooled, as shown in Fig. 1.10. In this way the beam path does not cross the turbulent cooling flow, so no optical aberration occur. On the other hand, this arrangement allows for less favorable surface/volume ratio for cooling and requires a more complex optical geometry.

Details of the amplifiers in the transmission geometry have been described in [58]. Both reflection and transmission geometry are still under consideration on the basis

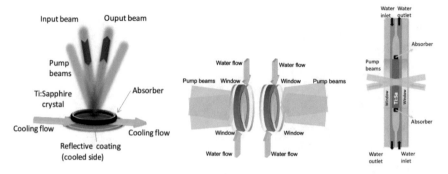

Fig. 1.10 Investigated geometries of an amplifier head for kW scale Ti:Sa average power laser, with disk gain material and water cooling. Left: Reflection geometry. Right: Transmission geometry [58]

of theoretical consideration and simulations, and both modelling and experimental investigations are needed.

In the mean time, other technologies are being developed which aim at even more efficient configurations removing the double-step in the pumping architecture, with direct pumping of the suitable ultrafast amplifier with diodes. Several direct CPA concepts have been explored in detail and some issues have emerged so far, including the minimum achievable pulse duration and the possibility of scaling to large size amplifiers. Examples of Direct CPA are Yb based systems like Yb:YAG and Yb:CaF_2. Compared to Ti:Sapphire, these media have allow direct pumping by semiconductor lasers, in the wavelength region 930–970 nm, without further wavelength conversion stages and exhibit a low quantum defect between the pump photon energy and the laser photon energy (around 10%) because the emission wavelength (usually 1030–1050 nm, depending on the host) is close to the pumping wavelength. This reduces the thermal load on the gain medium and thus the power dissipation requirements. Both these elements are advantageous in view of a high average power operation regime. Also, several hosts allow doping with Yb, providing flexibility in the choice of the gain media parameters like emission spectrum, thermal conductivity and thermo-optical parameters and the possibility of using ceramics technologies for large size gain elements. The main drawbacks of the Yb-based gain media are the reduced gain bandwidth, that makes it difficult to achieve pulse duration of 100 fs or less for high energy pulses and the high saturation fluence, often exceeding the damage threshold and preventing operation of the amplifiers in saturation.

Another direct CPA system is based on the Tm:YLF gain media that, when operated in the multi-pulse extraction, becomes as efficient as >70% at rep-rates >1 kHz, while at 100 Hz is still capable of approximately 20% efficiency [60]. Among the relevant host materials, YLF offers several attractive properties, including a negative dn/dT, low linear and nonlinear refractive indices, and natural birefringence. The Tm dopant in YLF emits laser radiation at ≈1.9 μm, and has a long upper-state lifetime (15 ms). It can be pumped with commercially-available, high-brightness continuous-wave laser diodes that operate at 800 nm CW pumping. Tm:YLF systems are being explored for future operation with 300 kW average power at as high as 10 kHz repetition rates for laser driven plasma accelerators [29].

In the long term, a very promising approach is based on Fiber laser technology, which is currently offering the best wall-plug efficiency for a laser, now exceeding 50% in CW mode. For high peak power architectures, solutions based on the coherent combination of a very large number of fiber amplifiers is being developed and prototyping is in progress. Studies show that this technology will be particularly suited and cost-competitive if laser driver parameters are going to evolve towards small energy (J level) per pulse and higher repetition rate (> 1 kHz).

In view of the above scenario of laser technologies it is clear that while major R&D effort is required, suitable laser driver solutions for laser-plasma accelerators can be conceived to meet requirements at different scales, from the single stage, table-top devices to future multi-stage colliders [61].

1.6 Industrial Development of Laser Drivers

High power laser science and technology R&D has always been inherently related to industrial developments. Large scientific endeavours like the inertial confinement fusion (ICF) research, have generated a strong impulse in the development of industrial products related to long-pulse (ns) technology, With the advent of ultrashort CPA pulses, industry has been mainly attracted by relatively small (TW scale) scale systems, with only a few companies focusing on multi TW systems and now aiming at multi-PW scale, mainly for scientific purposes. The potential use of such systems as power drivers for novel particle accelerators is setting new conditions, stimulating industrial investments towards higher repetition rate systems. Indeed, kW scale lasers suitable for pumping Ti:Sa in the 10–20 Hz range repetition rate are just emerging from industry systems like the P60 [62] or proto-types like the DIPOLE [63], and can be integrated in advanced Ti:Sa amplifiers design, provided a geometry with efficient cooling ensures heat removal from the amplifier head, as discussed in the preceding section. This is an important conceptual aspect of laser design that has impact on both the complexity and the compactness of the final system. Higher repetition rates will require significant technology developments to increase the repetition rate of currently available high power diode lasers and to enhance thermal management in the amplifier head and major numerical modelling and experimental data will be needed to demonstrate the path to commercial availability of a reliable kHz laser driver.

On the other hand, practical laser-plasma accelerator schemes include a number of components that ensure control and stability of the wakefield generation. The laser driver stability plays a key role here, with focal spot intensity distribution, energy and pointing stability, pulse duration etc. being closely related to the stability of the accelerated electron bunch. In particular, requirements on beam pointing stability are highly demanding, with required pointing accuracy well below the μrad to ensure no impact on the pointing instability of the accelerated electron bunch, typically set by the acceleration process. Here a combination of pointing detector and active pointing control is envisaged in the full scale implementation of the driver, calling for outstanding engineered industrial solutions.

These are just examples of the strong impact that laser-plasma accelerators are expected to have on the industrial development of lasers and opto-mechanical components to enable the expected specifications, further developing concepts currently under exploration at laser labs.

1.7   Multidisciplinary Applications

Laser-plasma acceleration (LPA) is increasingly being considered for the development of novel radiation sources and applications to industrial, medical and biological sciences. In the context of electron acceleration, given the ever increasing level of control and reliability of available schemes, compact, laser-driven accelerators are being explored for radiotherapy and diagnostics applications in areas where electron beams with energy up to several tens of MeV are normally used as primary beams. A laser-driven electron accelerator may have several advantages compared to conventional linacs, ranging from the small size of the acceleration region, to the possibility of multiplexing the electron source using a single laser driver. From the point of view of the specific properties of laser accelerated electrons, given the ultra-short duration of laser accelerated bunches compared to conventional RF linacs, a new regime of ultrafast radiation biology is emerging [64]. One of the key aspects to be investigated here is the very short bunch length, typical of LPA electron bunches, that leads to ultra-high instantaneous dose-rate, orders of magnitude higher than conventional sources. In view of this, pre-clinical studies are needed to address the radiobiological effectiveness of laser-driven electron sources compared to conventional linacs used in medical applications, with a particular attention to the intra-operatory radiation therapy (IORT) [65, 66]. More recently, other regimes of radio-therapy, including the so-called FLASH radiotherapy [67] or the very high energy electron radiotherapy (VHEE) [68] are also being considered as possible effective applications of laser-driven beams.

Since the original demonstration [69], laser-driven ion acceleration has also been driving dedicated biological and medical application aimed at exploring new concepts of such novel compact accelerators while validating radiobiological effectiveness of laser-driven ion sources. Also, special attention is being dedicated to the development of customised beam-lines capable of exploiting the full potential of these sources, while delivering control of the main beam parameters at the ion-target interaction point (see Lecture by P. Cirrone). Several of such dedicated beam-lines have been established at laser laboratories and facilities, including the LIGHT project at the PHELIX laser [70] or similar initiatives at the J-Karen laser [71] or at the INO-ILIL-PW laser [72], and the ELIMAIA beamline at ELI-Beamlines [73] just to mention a few. It is worth stressing here that most of these beamlines rely on the established Target Normal Sheath acceleration mechanism first observed in [69], using a combination of laser enhancements like the high temporal contrast option, and properly designed targets, to optimize ion source maximum energy and flux. At the same time, new mechanisms have been identified to further boost the ion energy and possibly control the energy distribution of accelerated electrons. A wealth of recent experiments are gradually approaching demonstration of these advanced acceleration mechanisms. These topics were also addressed during the school and a dedicated lecture is included in these proceedings (see Lectures by M. Borghesi).

Other applications can take full advantage of the brightness and the micrometer-scale transverse source size and enable a dramatic increase of resolution in X-ray based imaging techniques and, in particular, phase-contrast imaging. Indeed, phase-contrast X-ray imaging is gaining strong interest in a variety of fields, including biomedical imaging (Wenz 2015), due to its capability of highlighting subtle details (Schults 2010) at soft tissue interfaces without (or with minimal use) of exogenous contrast media. Both planar and tomographic applications have been devised so far, including in-vivo (preclinical) and ex-vivo imaging, in various disease models ranging from oncology (e.g., breast imaging), to neurology and cardiovascular imaging (Bravin 2013). Both propagation-based or gating-based phase-contrast imaging is being investigated. In the first case, a very small focal spot size is required (of the order of few micrometers) and long source-object distances must be used in order to spatially resolve tissue-driven X-ray phase shifts with common X-ray detectors. Due to the great potential of this imaging modality to overcome known limitations of current absorption-based X-ray imaging, effort is needed in order to translate phase-contrast imaging from the bench (preclinical) to the bedside (clinical) in the shortest time. Unfortunately, phase-contrast imaging requires high brightness spatially coherent sources that are not readily available at a laboratory scale. Apart from the small number of synchrotron sources that are used only for proof of principle imaging experiments, standard laboratory X-ray source technology is unable to provide the required source size and brilliance, thus requiring long exposure times and complex deconvolution, strongly limiting the clinical development of this outstanding imaging technique.

As discussed here, laser-plasma sources are capable of driving X-ray emission with the required degree of coherence and brilliance in a compact footprint and, with currently evolving laser technology, are bound to become an excellent solution to enable phase contrast imaging to emerge from the laboratory and become an industrial source for diffuse medical use. Developments are required to enhance the stability of the source and current laboratory effort focuses on this aspect, with innovative and original acceleration concepts that enable control of the main parameters, including source size and energy, using reliable temporal and spectral laser manipulation.

1.8 Summary

This introductory Chapter was conceived to give readers a preview of the topics presented during the School and discussed in depth in the following Chapters of this book. The underlying theme is the development of novel particle accelerators based on some of the most powerful physical mechanisms that can be activated using ultra-high intensity laser and their interaction with plasmas.

References

1. J. Faure, Y. Glinec, A. Pukhov, S. Kiselev, S. Gordienko, E. Lefebvre, J. Rousseau, F. Burgy, V. Malka, Monoenergetic beams of relativistic electrons from intense laser-plasma interactions. Nature **431**, 541 (2004)
2. C.G.R. Geddes, Cs. Toth, J. van Tilborg, E. Esarey, C.B. Schroeder, D. Bruhwiler, C. Nieter, J. Cary, and W.P. Leemans, Monoenergetic beams of relativistic electrons from intense laser–plasma interactions. Nature **431**, 538 (2004)
3. S.P.D. Mangles, C.D. Murphy, Z. Najmudin, A.G.R. Thomas, J.L. Collier, A.E. Dangor, E.J. Divall, P.S. Foster, J.G. Gallacher, C.J. Hooker, D.A. Jaroszynski, A.J. Langley, W.B. Mori, P.A. Norreys, F.S. Tsung, R. Viskup, B.R. Walton, K. Krushelnick, Monoenergetic beams of relativistic electrons from intense laser-plasma interactions. Nature **431**, 535 (2004)
4. A.J. Gonsalves, K. Nakamura, J. Daniels, C. Benedetti, C. Pieronek, T.C.H. De Raadt, S. Steinke, J.H. Bin, S.S. Bulanov, J. Van Tilborg, C.G.R. Geddes, C.B. Schroeder, Cs. Tóth, E. Esarey, K. Swanson, L. Fan-Chiang, G. Bagdasarov, N. Bobrova, V. Gasilov, G. Korn, P. Sasorov, W.P. Leemans, Petawatt laser guiding and electron beam acceleration to 8 GeV in a laser-heated capillary discharge waveguide. Phys. Rev. Lett. **122**(8), 84801 (2019)
5. S. Steinke, J. Van Tilborg, C. Benedetti, C.G.R. Geddes, J. Daniels, K.K. Swanson, A.J. Gonsalves, K. Nakamura, B.H. Shaw, C.B. Schroeder, E. Esarey, W.P. Leemans, Staging of laser-plasma accelerators. Phys. Plasmas **23**(5), 3–8 (2016)
6. P. Tomassini, S. De Nicola, L. Labate, P. Londrillo, R. Fedele, D. Terzani, L.A. Gizzi, The resonant multi-pulse ionization injection. Phys. Plasmas **24**(10) (2017)
7. A. Giulietti, N. Bourgeois, T. Ceccotti, X. Davoine, S. Dobosz, P. D'Oliveira, M. Galimberti, J. Galy, A. Gamucci, D. Giulietti, L.A. Gizzi, D.J. Hamilton, E. Lefebvre, L. Labate, J.R. Marques, P. Monot, H. Popescu, F. Reau, G. Sarri, P. Tomassini, P. Martin, P.D. Oliveira, Intense gamma-ray source in the giant-dipole-resonance range driven by 10-TW laser pulses. Phys. Rev. Lett. (3–6), 105002 (2008)
8. G. Sarri, K. Poder, J.M. Cole, W. Schumaker, A. Di Piazza, B. Reville, T. Dzelzainis, D. Doria, L.A. Gizzi, G. Grittani, S. Kar, C.H. Keitel, K. Krushelnick, S. Kuschel, S.P.D. Mangles, Z. Najmudin, N. Shukla, L.O. Silva, D. Symes, A.G.R. Thomas, M. Vargas, J. Vieira, M. Zepf, Generation of neutral and high-density electron-positron pair plasmas in the laboratory. Nat. Commun. **6**, 1–8 (2015)
9. G. Sarri, D.J. Corvan, W. Schumaker, J.M. Cole, A. Di Piazza, H. Ahmed, C. Harvey, C.H. Keitel, K. Krushelnick, S.P.D. Mangles, Z. Najmudin, D. Symes, A.G.R. Thomas, M. Yeung, Z. Zhao, M. Zepf, Ultrahigh Brilliance Multi-MeV γ-Ray Beams from Nonlinear Relativistic Thomson Scattering. Phys. Rev. Lett. **113**(22), 224801 (2014)
10. J. Van Tilborg, S. Steinke, C.G.R. Geddes, N.H. Matlis, B.H. Shaw, A.J. Gonsalves, J.V. Huijts, K. Nakamura, J. Daniels, C.B. Schroeder, C. Benedetti, E. Esarey, S.S. Bulanov, N.A. Bobrova, P.V. Sasorov, W.P. Leemans, Active plasma lensing for relativistic laser-plasma-accelerated electron beams. Phys. Rev. Lett. **115**(18), 1–5 (2015)
11. T. Tajima, J.M. Dawson, Laser electron accelerator. Phys. Rev. Lett. **43**(4), 267 (1979)

12. A. Pukhov, J. Meyer ter Vehn, Laser wake field acceleration: the highly non-linear broken-wave regime. Appl. Phys. B **74**, 355–361 (2002)
13. S. Gordienko, A. Pukhov, Phys. Plasmas **12**, 043109 (2005)
14. B.B. Pollock, C.E. Clayton, J.E. Ralph, F. Albert, A. Davidson, L. Divol, C. Filip, K. Herpoldt, S.H. Glenzer, W. Lu, K.A. Marsh, J. Meinecke, W.B. Mori, A. Pak, T.C. Rensink, J.S. Ross, J. Shaw, G.R. Tynan, C. Joshi, D.H. Froula, Demonstration of a narrow energy spread ... Phys. Rev. Lett. **107**, 045001 (2011)
15. S.V. Bulanov, F. Pegoraro, A.M. Pukhov, A.S. Sakharov, Transverse-wake wave breaking. Phys. Rev. Lett. **78**(22), 4205–4208 (1997)
16. S. Bulanov, N. Naumova, F. Pegoraro, J. Sakai, Particle injection into the wave acceleration phase due to nonlinear wake wave breaking. Phys. Rev. E **58**(5), 5257–5260 (1995)
17. P. Tomassini, M. Galimberti, A. Giulietti, D. Giulietti, L.A. Gizzi, L. Labate, F. Pegoraro, Production of high-quality electron beams in numerical experiments of laser wakefield acceleration with longitudinal wave breaking. Phys. Rev. ST Accel. Beams **6**, 121301 (2003)
18. A.J. Gonsalves, K. Nakamura, C. Lin, D. Panasenko, S. Shiraishi, T. Sokollik, C. Benedetti, C.B. Schroeder, C.G.R. Geddes, J. van Tilborg, J. Osterhoff, E. Esarey, C. Toth, W.P. Leemans, Tunable laser plasma accelerator based on longitudinal density tailoring. Nat. Phys. **7**(11), 862–866 (2011)
19. A. Buck, J. Wenz, J. Xu, K. Khrennikov, K. Schmid, M. Heigoldt, J.M. Mikhailova, M. Geissler, B. Shen, F. Krausz, S. Karsch, L. Veisz, Shock-front injector for high-quality laser-plasma acceleration. Phys. Rev. Lett. **110**(18), 185006 (2013)
20. D. Umstadter, J.K. Kim, E. Dodd, Laser injection of ultrashort electron pulses into wakefield plasma waves. Phys. Rev. Lett. **76**, 2073 (1996)
21. E. Esarey, R.F. Hubbard, W.P. Leemans, A. Ting, P. Sprangle, Phys. Rev. Lett. **79**, 2682 (1997)
22. M. Chen, Z.M. Sheng, Y.Y. Ma, J. Zhang, Electron injection and trapping in a laser wakefield by field ionization to high-charge states of gases. J. Appl. Phys. **99**(5), 2004–2007 (2006)
23. L.L. Yu, E. Esarey, C.B. Schroeder, J.L. Vay, C. Benedetti, C.G.R. Geddes, M. Chen, W.P. Leemans, Two-color laser-ionization injection. Phys. Rev. Lett. **112**(12), 1–5 (2013)
24. P. Tomassini, S. De Nicola, L. Labate, P. Londrillo, R. Fedele, D. Terzani, F. Nguyen, G. Vantaggiato, L.A. Gizzi, High-quality GeV-scale electron bunches with the resonant multi-pulse ionization injection. Nucl. Instruments Methods Phys. Res. Sect. A Accel. Spectrometers Detect. Assoc. Equip. **909**, 1–4 (2018)
25. Fernando Brandi and Leonida Antonio Gizzi, Optical diagnostics for density measurement in high-quality laser-plasma electron accelerators. High Power Laser Sci. Eng. **7**, e26 (2019)
26. L.A. Gizzi, A. Giulietti, D. Giulietti, T. Afshar-Rad, V. Biancalana, P. Chessa, C. Danson, E. Schifano, S.M. Viana, O. Willi, Characterization of laser plasmas for interaction studies. Phys. Rev. E **49**(6), 5628–5643 (1994)
27. A. Giulietti, M. Lucchesi, M. Vaselli, Converging shock on laser plasma: density profiles by holografic interferometry. Opt. Commun. **47**(2), 131–136 (1983)
28. P. Tomassini, A. Giulietti, L.A. Gizzi, M. Galimberti, D. Giulietti, M. Borghesi, O. Willi, Analyzing laser plasma interferograms with a continuous wavelet transform ridge extraction technique: the method. Appl. Opt. **40**(35), 6561–6568 (2001)
29. P.A. Walker, P.D. Alesini, A.S. Alexandrova, M.P. Anania, N.E. Andreev, I. Andriyash, A. Aschikhin, R.W. Assmann, T. Audet, A. Bacci, I.F. Barna, A. Beaton, A. Beck, A. Beluze, A. Bernhard, S. Bielawski, F.G. Bisesto, J. Boedewadt, F. Brandi, O. Bringer, R. Brinkmann, E. Bründermann, M. Büscher, M. Bussmann, G.C. Bussolino, A. Chance, J.C. Chanteloup, M. Chen, E. Chiadroni, A. Cianchi, J. Clarke, J. Cole, M.E. Couprie, M. Croia, B. Cros, J. Dale, G. Dattoli, N. Delerue, O. Delferriere, P. Delinikolas, J. Dias, U. Dorda, K. Ertel, A. Ferran Pousa, M. Ferrario, F. Filippi, R. Fiorito, R.A. Fonseca, M. Galimberti, A. Gallo, D. Garzella, P. Gastinel, D. Giove, A. Giribono, L.A. Gizzi, F.J. Grüner, A.F. Habib, L.C. Haefner, T. Heinemann, B. Hidding, B.J. Holzer, S.M. Hooker, T. Hosokai, A. Irman, D.A. Jaroszynski, S. Jaster-Merz, C. Joshi, M.C. Kaluza, M. Kando, O.S. Karger, S. Karsch, E. Khazanov, D. Khikhlukha, A. Knetsch, D. Kocon, P. Koester, O. Kononenko, G. Korn, I. Kostyukov, L. Labate, C. Lechner, W.P. Leemans, A. Lehrach, F.Y. Li, X. Li, V. Libov, A. Lifschitz, V.

Litvinenko, W. Lu, A.R. Maier, V. Malka, G.G. Manahan, S.P.D. Mangles, B. Marchetti, A. Marocchino, A. Martinez De La Ossa, J.L. Martins, F. Massimo, F. Mathieu, G. Maynard, T.J. Mehrling, A.Y. Molodozhentsev, A. Mosnier, A. Mostacci, A.S. Mueller, Z. Najmudin, P.A.P. Nghiem, F. Nguyen, P. Niknejadi, J. Osterhoff, D. Papadopoulos, B. Patrizi, R. Pattathil, V. Petrillo, M.A. Pocsai, K. Poder, R. Pompili, L. Pribyl, D. Pugacheva, S. Romeo, A.R. Rossi, E. Roussel, A.A. Sahai, P. Scherkl, U. Schramm, C.B. Schroeder, J. Schwindling, J. Scifo, L. Serafini, Z.M. Sheng, L.O. Silva, T. Silva, C. Simon, U. Sinha, A. Specka, M.J.V. Streeter, E.N. Svystun, D. Symes, C. Szwaj, G. Tauscher, A.G.R. Thomas, N. Thompson, G. Toci, P. Tomassini, C. Vaccarezza, M. Vannini, J.M. Vieira, F. Villa, C.G. Wahlström, R. Walczak, M.K. Weikum, C.P. Welsch, C. Wiemann, J. Wolfenden, G. Xia, M. Yabashi, L. Yu, J. Zhu, A. Zigler. Horizon 2020 EuPRAXIA design study. J. Phys. Conf. Ser. **874** (2017)

30. L.A. Gizzi, D. Giulietti, A. Giulietti, P. Audebert, S. Bastiani, J.P. Geindre, A. Mysyrowicz, Simultaneous measurements of hard X-rays and second-harmonic emission in fs laser-target interactions. Phys. Rev. Lett. **76**(13), 2278 (1996)

31. L.A. Gizzi, M. Galimberti, A. Giulietti, D. Giulietti, M. Borghesi, H.D. Campbell, A. Schiavi, O. Willi, Relativistic laser interactions with preformed plasma channels and gamma-ray measurements. Laser Part. Beams **19**, 181 (2001)

32. D. Giulietti, M. Galimberti, A. Giulietti, L.A. Gizzi, M. Borghesi, P. Balcou, A. Rousse, J.P. Rousseau, High-energy electron beam production by femtosecond laser interactions with exploding-foil plasmas. Phys. Rev. E **64**, 015402(R) (2001)

33. A. Giulietti, N. Bourgeois, T. Ceccotti, X. Davoine, S. Dobosz, P. D'Oliveira, M. Galimberti, J. Galy, A. Gamucci, D. Giulietti, L.A. Gizzi, D.J. Hamilton, E. Lefebvre, L. Labate, J.R. Marquès, P. Monot, H. Popescu, F. Réau, G. Sarri, P. Tomassini, P. Martin, Intense gamma-ray source in the giant-dipole-resonance range driven by 10-tw laser pulses. Phys. Rev. Lett. **101**, 105002 (2008)

34. G. Sarri, J. Corvan, D.W. Schumaker, J.M. Cole, A. Di Piazza, H. Ahmed, C. Harvey, H. Keitel, C.K. Krushelnick, S.P.D. Mangles, Z. Najmudin, D. Symes, A.G.R. Thomas, M. Yeung, Z. Zhao, M. Zepf, Ultrahigh brilliance multi-mev γ-ray beams from nonlinear relativistic thomson scattering. Phys. Rev. Lett. **113**, 224801 (2014)

35. A. Ben-Ismaïl, J. Faure, V. Malka, Optimization of gamma-ray beams produced by a laser-plasma accelerator. Nucl. Instruments Methods Phys. Res. Sect. A Accel. Spectrometers, Detect. Assoc. Equip. **629**(1), 382–386 (2011)

36. A. Ben-Ismaïl, O. Lundh, C. Rechatin, J.K. Lim, J. Faure, S. Corde, V. Malka, Compact and high-quality gamma-ray source applied to 10 μm-range resolution radiography. Appl. Phys. Lett. **98**(26), 264101 (2011)

37. S. Corde, K. Ta Phuoc, G. Lambert, R. Fitour, V. Malka, A. Rousse, A. Beck, E. Lefebvre. Femtosecond x rays from laser-plasma accelerators. Rev. Mod. Phys. **85**(1), 1–48 (2013)

38. J. Wenz, S. Schleede, K. Khrennikov, M. Bech, P. Thibault, M. Heigoldt, F. Pfeiffer, S. Karsch, Quantitative X-ray phase-contrast microtomography from a compact laser-driven betatron source. Nat. Commun. **6**(May), 7568 (2015)

39. S. Kneip, C. McGuffey, F. Dollar, M.S. Bloom, V. Chvykov, G. Kalintchenko, K. Krushelnick, A. Maksimchuk, S.P.D. Mangles, T. Matsuoka, Z. Najmudin, C.A.J. Palmer, J. Schreiber, W. Schumaker, A.G.R. Thomas, V. Yanovsky, X-ray phase contrast imaging of biological specimens with femtosecond pulses of betatron radiation from a compact laser plasma wakefield accelerator. Appl. Phys. Lett. **99**(9), 093701 (2011)

40. Z. Najmudin, S. Kneip, M.S. Bloom, S.P.D. Mangles, O. Chekhlov, A.E. Dangor, A. Döpp, K. Ertel, S.J. Hawkes, J. Holloway, C.J. Hooker, J. Jiang, N.C. Lopes, H. Nakamura, P.A. Norreys, P.P. Rajeev, C. Russo, M.J.V. Streeter, D.R. Symes, M. Wing, Compact laser accelerators for X-ray phase-contrast imaging. Philos. Trans. A. Math. Phys. Eng. Sci. **372**, 20130032 (2014)

41. S. Cipiccia, M.R. Islam, B. Ersfeld, R.P. Shanks, E. Brunetti, G. Vieux, X. Yang, R.C. Issac, S.M. Wiggins, G.H. Welsh, M.-P. Anania, D. Maneuski, R. Montgomery, G. Smith, M. Hoek, D.J. Hamilton, N.R.C. Lemos, D. Symes, P.P. Rajeev, V.O. Shea, J.M. Dias, D.A. Jaroszynski,

Gamma-rays from harmonically resonant betatron oscillations in a plasma wake. Nat. Phys. **7**(11), 867–871 (2011)

42. R.H. Milburn, Electron scattering by an intense polarized photon field. Phys. Rev. Lett. **10**, 75–77 (1963)

43. C. Bemporad, R.H. Milburn, N. Tanaka, M. Fotino, High-energy photons from compton scattering of light on 6.0-Gev electrons. Phys. Rev. **138**, B1546–B1549 (1965)

44. P. Tomassini, A. Giulietti, D. Giulietti, L.A. Gizzi. Thomson backscattering X-rays from ultrarelativistic electron bunches and temporally shaped laser pulses. Appl. Phys. B **80**(4–5), 419–436 (2005)

45. S.K. Ride, E. Esarey, M. Baine, Thomson scattering of intense lasers from electron beams at arbitrary interaction angles. Phys. Rev. E **52**, 5425–5442 (1995)

46. W.P. Leemans, Stimulated compton scattering from preformed underdense plasmas. Phys. Rev. Lett. **67**(11), 1434–1437 (1991)

47. P. Tomassini, M. Galimberti, A. Giulietti, D. Giulietti, L.A. Gizzi, L. Labate, F. Pegoraro, Production of high-quality electron beams in numerical experiments of laser wakefield acceleration with longitudinal wave breaking. Phys. Rev. Spec. Top. Accel. Beams **6**(12), 121301 (2003)

48. E. Esarey, S.K. Ride, P. Sprangle, Nonlinear Thomson scattering of intense laser pulses from beams and plasmas. Phys. Rev. E **48**(4), 3003–3021 (1993)

49. H. Schwoerer, B. Liesfeld, H.-P. Schlenvoigt, K.-U. Amthor, R. Sauerbrey, Thomson-backscattered x rays from laser-accelerated electrons. Phys. Rev. Lett. **96**, 014802 (2006)

50. S. Chen, N.D. Powers, I. Ghebregziabher, C.M. Maharjan, C. Liu, G. Golovin, S. Banerjee, J. Zhang, N. Cunningham, A. Moorti, S. Clarke, S. Pozzi, D.P. Umstadter, MeV-energy x rays from inverse compton scattering with laser-wakefield accelerated electrons. Phys. Rev. Lett. **155003**(April), 1–5 (2013)

51. C. Liu, G. Golovin, S. Chen, J. Zhang, B. Zhao, D. Haden, S. Banerjee, J. Silano, H. Karwowski, D. Umstadter, Generation of 9 MeV γ-rays by all-laser-driven Compton scattering with second-harmonic laser light. Opt. Lett. **39**(14), 4132 (2014)

52. N.D. Powers, I. Ghebregziabher, G. Golovin, C. Liu, S. Chen, S. Banerjee, J. Zhang, D.P. Umstadter, Quasi-monoenergetic and tunable X-rays from a laser-driven Compton light source. Nat. Photonics **8**(1), 28–31 (2013)

53. K. Khrennikov, J. Wenz, A. Buck, J. Xu, M. Heigoldt, L. Veisz, S. Karsch, Tunable all-optical quasimonochromatic thomson x-ray source in the nonlinear regime. Phys. Rev. Lett. **114**(19), 1–5 (2015)

54. T.H. Maiman, Stimulated optical radiation in ruby. Nature **187**(4736), 493–494 (1960)

55. C. Danson et al., Petawatt class lasers worldwide. High Power Laser Sci. Eng. **7**, E54 (2019). https://doi.org/10.1017/hpl.2019.36

56. D. Strickland, G. Mourou, Compression of amplified chirped optical pulses. Opt. Commun. **56**(3), 219–221 (1985)

57. E. Sistrunk, T. Spinka, A. Bayramian, S. Betts, R. Bopp, S. Buck, K. Charron, J. Cupal, R. Deri, M. Drouin, A. Erlandson, E.S. Fulkerson, J. Horner, J. Horacek, J. Jarboe, K. Kasl, D. Kim, E. Koh, L. Koubikova, R. Lanning, W. Maranville, C. Marshall, D. Mason, J. Menapace, P. Miller, P. Mazurek, A. Naylon, J. Novak, D. Peceli, P. Rosso, K. Schaffers, D. Smith, J. Stanley, R. Steele, S. Telford, J. Thoma, D. VanBlarcom, J. Weiss, P. Wegner, B. Rus, C. Haefner, All diode-pumped, high-repetition-rate advanced petawatt laser system (HAPLS) (2017)

58. L.A. Gizzi, P. Koester, L. Labate, F. Mathieu, Z. Mazzotta, G. Toci, M. Vannini, A viable laser driver for a user plasma accelerator. Nucl. Instruments Methods Phys. Res. Sect. A Accel. Spectrometers, Detect. Assoc. Equip. **909**, 58–66 (2018)

59. See web-site of "the international committee on ultra-high intensity lasers", https://www.icuil.org/

60. T.C. Galvin et al., Scaling of petawatt-class lasers to multi-kHZ repetition rates. Proc. SPIE **11033** (2019)

61. C.B. Schroeder, C. Benedetti, E. Esarey, W.P. Leemans, Laser-plasma-based linear collider using hollow plasma channels. Nucl. Instruments Methods Phys. Res. Sect. A Accel. Spectrometers, Detect. Assoc. Equip. (2016)

62. S. Kühn, M. Dumergue, S. Kahaly, S. Mondal, M. Füle, T. Csizmadia, B. Farkas, B. Major, Z. Várallyay, F. Calegari, M. Devetta, F. Frassetto, E. Månsson, L. Poletto, S. Stagira, C. Vozzi, M. Nisoli, P. Rudawski, S. Maclot, F. Campi, H. Wikmark, C.L. Arnold, C.M. Heyl, P. Johnsson, A. L'Huillier, R. Lopez-Martens, S. Haessler, M. Bocoum, F. Boehle, A. Vernier, G. Iaquaniello, E. Skantzakis, N. Papadakis, C. Kalpouzos, P. Tzallas, F. Lépine, D. Charalambidis, K. Varjú, K. Osvay, G. Sansone, The ELI-ALPS facility: the next generation of attosecond sources. J. Phys. B At. Mol. Opt. Phys. **50**(13), 132002 (2017)

63. P. Mason, M. Divoký, K. Ertel, J. Pilař, T. Butcher, M. Hanuš, S. Banerjee, J. Phillips, J. Smith, M. De Vido, A. Lucianetti, C. Hernandez-Gomez, C. Edwards, T. Mocek, J. Collier, Kilowatt average power 100 J-level diode pumped solid state laser. Optica **4**(4), 438 (2017)

64. V. Malka, J. Faure, Y. Gaudel, Ultra-short electron beams based spatio-temporal radiation biology and radiotherapy. Mutat. Res. **704**, 142 (2010)

65. S. Righi, E. Karaj, G. Felici, F. Di Martino, Dosimetric characteristics of electron beams produced by two mobile accelerators, novac7 and liac, for intraoperative radiation therapy through monte carlo simulation. J. Appl. Clin. Med. Phys. **14**(1) (2013)

66. A. Gamucci, N. Bourgeois, T. Ceccotti, X. Davoine, S. Dobosz, P. D'Oliveira, M. Galimberti, J. Galy, A. Giulietti, D. Giulietti, L.A. Gizzi, D.J. Hamilton, L. Labate, E. Lefebvre, J.R. Marquès, P. Monot, H. Popescu, F. Réau, G. Sarri, P. Tomassini, P. Martin, Laser-IORT: A laser-driven source of relativistic electrons suitable for Intra-Operative Radiation Therapy of tumors, in AIP Conference Proceedings vol. 1209 (2010) pp. 39–42

67. M. Durante, E. Bräuer-Krisch, M. Hill, Faster and safer? FLASH ultra-high dose rate in radiotherapy. Br J Radiol **91**(1082), 20170628 (2018, Feb). https://doi.org/10.1259/bjr.20170628. Epub 2017, Dec 15

68. E. Schüler, K. Eriksson, E. Hynning, S.L. Hancock, S.M. Hiniker, M. Bazalova-Carter, T. Wong, Q.-T. Le, B.W. Loo Jr., P.G. Maxim, Very high-energy electron (VHEE) beams in radiation therapy; Treatment plan comparison between VHEE, VMAT, and PPBS. Med. Phys. **44**(6):2544–2555 (2017, Jun). https://doi.org/10.1002/mp.12233. Epub 2017, May 4

69. R.A. Snavely, M.H. Key, S.P. Hatchett, T.E. Cowan, M. Roth, T.W. Phillips, M.A. Stoyer, E.A. Henry, T.C. Sangster, M.S. Singh, S.C. Wilks, A. MacKinnon, A. Offenberger, D.M. Pennington, K. Yasuike, A.B. Langdon, B.F. Lasinski, J. Johnson, M.D. Perry, E.M. Campbell, Intense high-energy proton beams from petawatt-laser irradiation of solids. Phys. Rev. Lett. **85**(14), 2945–2948 (2000)

70. S. Busold, D. Schumacher, O. Deppert, C. Brabetz, F. Kroll, A. Blažević, V. Bagnoud, M. Roth, Commissioning of a compact laser-based proton beam line for high intensity bunches around 10 mev. Phys. Rev. ST Accel. Beams **17**, 031302 (2014)

71. A. Yogo, T. Maeda, T. Hori, H. Sakaki, K. Ogura, M. Nishiuchi, A. Sagisaka, H. Kiriyama, H. Okada, S. Kanazawa, T. Shimomura, Y. Nakai, M. Tanoue, F. Sasao, P.R. Bolton, M. Murakami, T. Nomura, S. Kawanishi, K. Kondo, Measurement of relative biological effectiveness of protons in human cancer cells using a laser-driven quasimonoenergetic proton beamline. Appl. Phys. Lett. **98**(5), 2–4 (2011)

72. L. Gizzi, D. Giove, C. Altana, F. Brandi, P. Cirrone, G. Cristoforetti, A. Fazzi, P. Ferrara, L. Fulgentini, P. Koester, L. Labate, G. Lanzalone, P. Londrillo, D. Mascali, A. Muoio, D. Palla, F. Schillaci, S. Sinigardi, S. Tudisco, G. Turchetti, A new line for laser-driven light ions acceleration and related TNSA studies. Appl. Sci. **7**(10), 984 (2017)

73. F. Romano, F. Schillaci, G.A.P. Cirrone, G. Cuttone, V. Scuderi, L. Allegra, The ELIMED transport and dosimetry beamline for laser-driven ion beams nuclear instruments and methods. Phys. Res. A. **829**, 153–158 (2016)

Chapter 2
Basics of Laser-Plasma Interaction: A Selection of Topics

Andrea Macchi

Abstract A short, tutorial introduction to some basic concepts of laser-plasma interactions at ultra-high intensities is given. The selected topics include (a) elements of the relativistic dynamics of an electron in electromagnetic fields, including the ponderomotives force and classical radiation friction; (b) the "relativistic" nonlinear optical transparency and self-focusing; (c) the moving mirror concept and its application to light sail acceleration and high harmonic generation, with a note on related instabilities; (d) some specific phenomena related to the absorption of energy, kinetic momentum and angular momentum from the laser light.

2.1 Introduction

Present-day short pulse, high power laser systems have reached the petawatt (10^{15} W) level. When such power is tightly focused in a spot with a diameter of few wavelengths λ ($\simeq 1$ μm for sub-picosecond systems), intensities exceeding 10^{21} W cm^{-2} may be achieved. The corresponding strength of the EM fields is such that any sample of matter exposed to such fields becomes instantaneously highly ionized, i.e. turned into a plasma, and the freed electrons oscillate with momenta largely exceeding $m_e c$ (where m_e is the electron mass and c is the speed of light). The nonlinear dynamics of such relativistic plasma in a superstrong EM field is the basis of advanced schemes of laser-plasma sources of high energy electrons, ions and photons which are characterized by high brilliance and ultrashort duration.

A few years ago we tried to present the basic concepts of the theory of superintense laser-plasma interactions in a primer of about one hundred of pages [1], and it is hard to further condensate such material. Thus, the present paper is mostly an ultrashort

A. Macchi (✉)
National Institute of Optics, National Research Council (CNR/INO), Adriano Gozzini laboratory, via Giuseppe Moruzzi 1, 56124 Pisa, Italy
e-mail: andrea.macchi@ino.cnr.it

Enrico Fermi Department of Physics, University of Pisa, largo Bruno Pontecorvo 3, 56127 Pisa, Italy

© Springer Nature Switzerland AG 2019

L. A. Gizzi et al. (eds.), *Laser-Driven Sources of High Energy Particles and Radiation*, Springer Proceedings in Physics 231, https://doi.org/10.1007/978-3-030-25850-4_2

introduction to the field at a "sub-primer" level, focused on an arbitrary selection of
contents. We do not enter into mathematical details which can be found in the primer
or in the other (few) references we cite.

Our rough selection criterion is to include here preferentially topics on which either we witnessed frequent misunderstanding or we may add something with respect
to our primer. Beyond the latter, more complete and advanced introductions may be
found in textbooks [2, 3] or review papers [4, 5]. We also address the reader to other
reviews for the important topics of laser-plasma accelerators of both electrons [6]
and ions [7], on which additional references may be found in other contributions to
this book. On topics where controversies are present, we have only room to give our
personal point of view.

2.2 Single Electron Dynamics and Radiation Friction

A look at the dynamics of a single electron in an EM field of arbitrary amplitude is a
good warm-up before discussing a many-particle system with collective effects, i.e.
a plasma. In non-covariant notation, the relativistic motion of an electron in a *given*
EM field is described by the equations

$$\frac{d\mathbf{p}}{dt} = -e\left(\mathbf{E} + \frac{\mathbf{v}}{c} \times \mathbf{B}\right) , \qquad \frac{d\mathbf{r}}{dt} = \mathbf{v} , \qquad \frac{d(m_e\gamma c^2)}{dt} = -e\mathbf{v} \cdot \mathbf{E} , \qquad (2.1)$$

where $\mathbf{p} = \mathbf{p}(t)$, $\mathbf{r} = \mathbf{r}(t)$, $\mathbf{v} = \mathbf{v}(t) = \mathbf{p}/m_e\gamma$, $\gamma = (1 + \mathbf{p}^2/m_e^2 c^2)^{1/2} = (1 - v^2/c^2)^{-1/2}$, and the fields are evaluated at the electron position, i.e. $\mathbf{E} = \mathbf{E}(\mathbf{r}(t), t)$
and $\mathbf{B} = \mathbf{B}(\mathbf{r}(t), t)$. By *given* fields we mean that we neglect their self-consistent
modification by the motion of the electron (see Sect. 2.2.3).

2.2.1 Motion in Plane Wave Fields

Exact relations and solutions can be found for plane wave fields, conveniently described by the vector potential $\mathbf{A} = \mathbf{A}(x - ct)$ which we take to be propagating along
$\hat{\mathbf{x}}$. The EM fields are given by $\mathbf{E} = -\partial_t\mathbf{A}/c$ and $\mathbf{B} = \nabla \times \mathbf{A} = \hat{\mathbf{x}} \times \partial_x\mathbf{A}$. By separating the electron momentum in longitudinal (p_x) and transverse ($\mathbf{p}_\perp$) components,
it is possible to find two constants of motion:

$$\frac{d}{dt}\left(\mathbf{p}_\perp - \frac{e}{c}\mathbf{A}\right) = 0 , \qquad \frac{d}{dt}(p_x - m_e\gamma c) = 0 . \qquad (2.2)$$

The first relation is the conservation of canonical momentum related to the traslational
invariance in the transverse plane (yz). The second arises from the properties of the
EM field: if a net amount of energy $\mathcal{E}$ is absorbed from the field, a proportional

amount of momentum $\mathcal{E}/c$ must be absorbed as well.[1] If an electron is initially at rest before it is reached by the wave, then $\mathbf{p}_\perp = e\mathbf{A}/c$ and $p_x = mc(\gamma - 1)$ at any time. These relations also yield $p_x = e^2\mathbf{A}^2/2m_ec^3$ and imply that, as the field is over ($\mathbf{A} = 0$), an electron initially at rest will be at rest again, i.e. no net acceleration is possible in a plane EM wave.

Now consider the case of a monochromatic wave of frequency ω,

$$\mathbf{A} = A_0 \left[\hat{\mathbf{y}} \cos\theta \cos(kx - \omega t) + \hat{\mathbf{z}} \sin\theta \sin(kx - \omega t) \right] , \qquad \mathbf{B} = \hat{\mathbf{x}} \times \mathbf{E}, \qquad (2.3)$$

where $k = \omega/c$ and $-\pi/2 < \theta < \pi/2$ determines the wave polarization: for instance $\theta = 0$ and $\theta = \pm\pi/2$ correspond to linear polarization (LP), while $\theta = \pm\pi/4$ corresponds to circular polarization (CP). This wave has infinite duration, but one may still assume the same initial conditions as above if the wave is "turned on" over an arbitrarily long rising time. One thus obtains an average drift momentum $\langle p_x \rangle = \langle e^2\mathbf{A}^2/2m_ec^3 \rangle$ (the brackets denote an average over a laser period). The trajectories (Fig. 2.1a–b) have a self-similar form, i.e. they can be written as function of the scaled coordinates x/a_0^2, y/a_0 and z/a_0 where a_0 is a dimensionless amplitude of the EM wave,

$$a_0 = \frac{eA_0}{m_ec^2} . \qquad (2.4)$$

The drift velocity is $v_D = ca_0^2/(a_0^2 + 4)$. By transforming to a frame moving with such velocity along $\hat{\mathbf{x}}$, the trajectories become closed. For LP the electron performs a "figure of eight" in the plane containing $\hat{\mathbf{x}}$ and the polarization direction (Fig. 2.1). For CP, the electron moves on a circle in the yz plane. Notice that in this latter case the γ-factor is a constant and the motion does not contain high harmonics of ω.

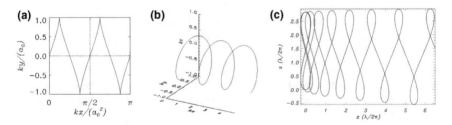

Fig. 2.1 a, b self-similar "drifting" trajectories of an electron in a monochromatic plane wave for linear (**a**) and circular (**b**) polarization. **c** the figure-of-eight trajectory (red line) obtained by subtracting the drift from case (**a**), and the trajectory with same initial conditions, but adding the radiation friction force (black line)

---

[1] In fact, in classical electrodynamics the ratio between the amount of energy and of momentum modulus in a wavepacket is c, thus this relation must be conserved if the wavepacket is totally absorbed by a medium. In a quantum picture, one may think of the absorption of a given number of photons, each having energy $\mathcal{E} = \hbar\omega$ and momentum modulus $\mathcal{E}/c$.

The parameter a_0 introduced in (2.4) is a convenient indicator of the onset of the relativistic dynamics regime. In the "no drift" frame, the typical value of the gamma factor (temporally averaged for LP) is $\gamma = (1 + a_0^2/2)^{1/2}$, thus the dynamics is strongly relativistic when $a_0 \gg 1$. The parameter is related to the wave intensity I and wavelength λ by $a_0 = 0.85 \, (I\lambda^2/10^{18} \, \text{Wcm}^{-2}\mu\text{m}^2)^{1/2}$.

2.2.2 Ponderomotive Force

The motion in a plane wave is an useful reference case, but in most cases we have to deal with more complex field distributions, such as a laser pulse with a finite extension in space and time. At least we may assume the field to be *quasi-monochromatic*, i.e. to be described by $\mathbf{A}(\mathbf{r}, t) = \text{Re}\left[\tilde{\mathbf{A}}(\mathbf{r}, t)e^{-i\omega t}\right]$ with $\langle \mathbf{A}(\mathbf{r}, t)\rangle \simeq 0$ and $\left\langle \tilde{\mathbf{A}}(\mathbf{r}, t)\right\rangle \simeq \tilde{\mathbf{A}}(\mathbf{r}, t)$, i.e. the envelope function $\tilde{\mathbf{A}}(\mathbf{r}, t)$ describes the temporal variation of the field on a scale slower than the oscillation at frequency ω. The idea is to separate these different scales by writing for the position $\mathbf{r}(t) \equiv \mathbf{r}_s(t) + \mathbf{r}_o(t)$ where $\langle \mathbf{r}_s(t)\rangle \simeq \mathbf{r}_s(t)$ and $\langle \mathbf{r}_o(t)\rangle \simeq 0$, i.e. $\mathbf{r}_o(t)$ describes the fast oscillation around the slowly-moving center $\mathbf{r}_s(t)$. In the non-relativistic case, one obtains equations for the "slow" motion as

$$m_e \frac{d\mathbf{v}_s}{dt} = -\frac{e^2}{2m_e\omega^2}\nabla \left\langle \mathbf{E}^2(\mathbf{r}_s(t), t)\right\rangle \equiv \mathbf{F}_p \,, \qquad \frac{d\mathbf{r}_s}{dt} = \mathbf{v}_s \,, \qquad (2.5)$$

where $\mathbf{F}_p$ is named the *ponderomotive* force (PF).[2] Equation (2.5) is based on a perturbative approach where magnetic effects are taken into account up to first order in v/c, and the spatial variation of the fields over a wavelength is small ($|\lambda\nabla E| \ll E$).

According to (2.5) the electrons are pushed out of the regions where the field is higher. Thus, if a laser pulse propagates through a tenuous plasma (Fig. 2.2), electrons will be pushed in the forward (propagation) direction on the leading edge of the pulse, and in the backward direction on the trailing edge: in proper conditions, this effect generates wake waves in the plasma [6]. The PF associated to the intensity gradient in the radial direction tends to pile electrons at the edge of the laser beam and create a low-density channel along the propagation path, which can cause a self-guiding effect (see Sect. 2.4.2).

An extension of the PF to the relativistic regime is not straightforward. For a quasi-transverse, quasi-plane wave field one may follow the hint that the non-relativistic PF (2.5) is the gradient of the average oscillation energy ("ponderomotive potential"). Assuming $\mathbf{p}_\perp \simeq e\mathbf{A}/c$ and $\gamma \simeq (1 + \mathbf{p}_\perp^2/m_e^2c^2)^{1/2}$, one can write the oscillation energy in the relativistic case as $m_e c^2(\gamma - 1)$ and replace the potential in (2.5). However,

---

[2]We stress that we define the PF as a cycle-averaged approximation of the Lorentz force. However, in the literature sometimes the term *"oscillating PF"* has been used [8] to refer to oscillating nonlinear terms in the Lorentz force (such as the $\mathbf{v} \times \mathbf{B}$ term which has a 2ω component). This definition is inconsistent with the whole idea of separating the "slow" and "fast" scales in the motion.

Fig. 2.2 Ponderomotive
scattering of electrons by the
ponderomotive force (2.5) of
a laser pulse having finite
length and width

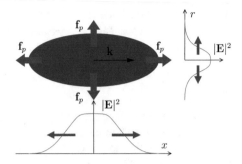

one has also to take into account that the oscillatory motion yields relativistic inertia.
One may thus write

$$\frac{d}{dt}(m_{\text{eff}}\mathbf{v}_s) \simeq -\nabla(m_{\text{eff}}c^2) , \qquad m_{\text{eff}} \equiv m_e(1 + \langle \mathbf{a}^2 \rangle (\mathbf{r}_s, t))^{1/2} , \qquad (2.6)$$

(where $\mathbf{a} = e\mathbf{A}/m_e c^2$) with m_{eff} acting as an effective, position- and time-dependent
mass. We remark that this expression is limited to a "semi-relativistic case", in
which the average velocity $|\mathbf{v}_s| \ll c$, and for smooth field profiles where transverse
components are much larger than longitudinal ones (e.g. a loosely focused laser
beam).

2.2.3   Radiation Friction (Reaction)

While an electron is accelerated by an EM field, it also radiates EM waves when
accelerated. But the "standard" equations of motion (2.1) do not account for the
energy and momentum carried away by the radiation. For example, according to
(2.1) an electron in an uniform and constant magnetic field performs a circular orbit
at constant energy; but since the electron experiences a centripetal acceleration, it
will radiate and lose energy, so that we expect the trajectory to become a spiral as
if the electron was experiencing a friction force. To describe such *radiation friction*
(RF) effects, additional terms must be added to the Lorentz force in order that the
motion is self-consistent with the radiation emission. The phenomenon can also be
described as the back-action of the fields generated by the electron on itself, so it is
also named *radiation reaction* (RR).

RR (or RF) is a longstanding and classic problem of classical electrodynamics.
In ordinary conditions the effect is either negligible or at least it can be treated
perturbatively and phenomenologically, e.g. inserting a simple friction force. The
dynamics of the electron becomes strongly affected by the radiation emission when
the energy of the emitted radiation is comparable to the work done on the electron by
the accelerating fields ([9], Sect. 16.1), which implies field strengths at the frontier of
those produced by present-day laser technology. This circumstance has revitalized

the debate (and associated controversy) on RR in recent years. However, it is apparent that as long as a classical description is adequate, one can safely use the RR force given in the textbook by Landau and Lifshitz (LL) [10]:

$$\mathbf{F}_{RR} \simeq -\frac{2r_c^2}{3} \left(\gamma^2 \left(\mathbf{L}^2 - \left(\frac{\mathbf{v}}{c} \cdot \mathbf{E} \right)^2 \right) \frac{\mathbf{v}}{c} - \mathbf{L} \times \mathbf{B} - \left(\frac{\mathbf{v}}{c} \cdot \mathbf{E} \right) \cdot \mathbf{E} \right), \qquad (2.7)$$

where $\mathbf{L} \equiv \mathbf{E} + \mathbf{v} \times \mathbf{B}/c$, $r_c = e^2/m_e c^2$ is the classical electron radius, and small terms containing the temporal derivatives of the fields have been dropped down [11]. It may be interesting to notice that for an electron which is instantaneously at rest ($\mathbf{v} = 0$) the force reduces to

$$\mathbf{F}_{RR} \simeq \frac{2r_c^2}{3} \mathbf{E} \times \mathbf{B} = \sigma_T \frac{\mathbf{S}}{c}, \qquad (2.8)$$

where $\sigma_T = 8\pi r_c^2/3$ is the Thomson cross section for the scattering of an EM wave, and $\mathbf{S} = c\mathbf{E} \times \mathbf{B}/4\pi$ is the Poynting vector giving the energy flux of the wave (the intensity $I = |\mathbf{S}|$): thus, in this limit the RR force is a drag force which describes the absorption of an amount of EM momentum proportional to the amount of EM energy subtracted from the wave and then radiated away.

An exact solution for the motion in a plane EM wave exists also when the RR force (2.7) is included [12]. The modification of the trajectory is shown in Fig. 2.1, for the same initial conditions yielding the closed "figure of eight" when neglecting RR: if the latter is included, the trajectory opens up with the electron gaining energy and accelerating along the propagation direction. Of course a friction force sounds as unable to accelerate anything, but actually the effect of friction is to change the relative phase between the fields and the electron velocity. This yields $\langle \mathbf{v} \cdot \mathbf{E} \rangle \neq 0$, so that the electron gains energy from the wave, and $\langle \mathbf{v} \times \mathbf{B} \rangle \neq 0$, so that the electron is accelerated along $\hat{\mathbf{x}}$.

The classical theory predicts that the spectrum of the radiation scattered from a relativistic electron peaks at frequencies $\omega_{rad} \simeq \gamma^3 \omega_i$ ([9], Sect. 14.4), where ω_i is the frequency of the incident radiation ($\omega_{rad} = \omega_i$ in the linear non-relativistic regime). Thus, with increasing γ eventually the energy of a single photon $\hbar\omega_{rad} \gtrsim m_e c^2 \gamma$, the electron energy, so that the recoil from the photon emission is not negligible and a quantum electrodynamics (QED) description becomes necessary. This is reminiscent of the well-known Compton scattering, but here the relevant regime involves the sequential absorption of very many low-frequency photons and the emission of several high-frequency photons. A QED theory of RR is still an open issue and is the subject of current research (see [13] for a discussion).

2.3   Kinetic and Fluid Equations

For a plasma of electrons and ions at high energy density, a classical approach is adequate. The most complete description of the dynamics is based on the knowledge of the distribution function $f_a = f_a(\mathbf{r}, \mathbf{p}, t)$ which gives the density of particles in the phase space $(\mathbf{r}, \mathbf{p})$ for all species a (e.g. $a = e, i$ for a single ion distribution).

A great simplification arises from the possibility of neglecting binary collisions, since the cross section for Coulomb scattering quickly decreases with increasing particle energy. For further simplicity we neglect any process which may create or destroy particles (such as ionization, pair production, ...), as well as radiation friction (RF) whose inclusion will be discussed later. The total number of particles of each species is thus conserved, and the distribution function satisfies a continuity equation in the phase space (the Vlasov equation):

$$\frac{\partial f_a}{\partial t} + \frac{\partial}{\partial \mathbf{r}}(\dot{\mathbf{r}}_a f_a) + \frac{\partial}{\partial \mathbf{p}}(\dot{\mathbf{p}}_a f_a) = 0 , \tag{2.9}$$

where

$$\dot{\mathbf{r}}_a = \mathbf{v} = \frac{\mathbf{p}c}{(\mathbf{p}^2 + m_a^2 c^2)^{1/2}} , \qquad \dot{\mathbf{p}}_a = q_a \left(\mathbf{E} + \frac{\mathbf{v}}{c} \times \mathbf{B} \right) . \tag{2.10}$$

The coupling with Maxwell equations for the EM fields $\mathbf{E} = \mathbf{E}(\mathbf{r}, t)$ and $\mathbf{B} = \mathbf{B}(\mathbf{r}, t)$ occurs via the charge and current densities obtained from f_a:

$$\rho(\mathbf{r}, t) = \sum_a q_a \int f_a \mathrm{d}^3 p , \qquad \mathbf{J}(\mathbf{r}, t) = \sum_a q_a \int \mathbf{v} f_a \mathrm{d}^3 p . \tag{2.11}$$

The Vlasov-Maxwell system constitutes the basis for the kinetic description of laser-plasma interactions, mostly via numerical simulations based on particle-in-cell (PIC) codes [14]. The PIC method may be extended to include collisions, ionization, and particle production (see e.g. [15, 16]). RF effects can be included straightforwardly by adding the LL force 2.7 (Sect. 2.2.3) to the second of (2.10).[3] The technical implementation in PIC codes proposed in [11] has been successfully benchmarked in [17]. Notice that in a simulation, because of the finite resolution of a spatial grid over which the fields are represented, it is almost impossible to resolve the high-energy radiation emitted by ultra-relativistic electrons at frequencies $\omega_{\mathrm{rad}} \simeq \gamma^3 \omega$, with ω the frequency of the driving lasers. However, radiation of such frequency escapes even from a solid-density plasma with negligible interactions, and it is of incoherent nature being of such small wavelength $\lambda_{\mathrm{rad}} = 2\pi c/\omega_{\mathrm{rad}}$ that $n_e \lambda_{\mathrm{rad}}^3 \ll 1$. Thus, RF losses in a laser-plasma interaction are simply measured by the amount of energy which "disappears" from the simulations.[4]

---

[3]Notice that in (2.9)–(2.10) $\partial_{\mathbf{r}}(\dot{\mathbf{r}}_a f_a) = \dot{\mathbf{r}}_a \partial_{\mathbf{r}} f_a$ and $\partial_{\mathbf{p}}(\dot{\mathbf{p}}_a f_a) = \dot{\mathbf{p}}_a \partial_{\mathbf{p}} f_a$, as it is usual to write for the Vlasov equation. However, if the LL force is added to the Lorentz force, $\partial_{\mathbf{p}}(\dot{\mathbf{p}}_a f_a) \neq \dot{\mathbf{p}}_a \partial_{\mathbf{p}} f_a$. This is not an issue for the standard PIC algorithms which provide a solution of the general kinetic equation (2.9).

[4]In principle also low-frequency, coherent radiation which is resolved in the simulation contributes to the RF effect, thus there is some double counting of such radiation in the force since it is included both in the Lorentz and in the LL terms. However, for highly relativistic electrons with $\gamma \gg 1$ the

While a kinetic approach is most of the times necessary for a comprehensive study of laser-plasma interaction phenomena, the simplified description based on moments of (2.9), i.e. on fluid equations, provides a suitable ground for basic models. As the motion of electrons is dominated by the superintense EM fields, one may neglect the "random" or thermal component of the motion and the associated pressure term, and obtain a closed set of moment equations. This is named the "cold" fluid approximation although the name might sound funny for such a high energy density plasma. Introducing the electron density $n_e = n_e(\mathbf{r}, t)$ and fluid momentum $\mathbf{p}_e = \mathbf{p}_e(\mathbf{r}, t)$,

$$n_e(\mathbf{r}, t) \equiv \int f_e d^3 p , \qquad \mathbf{p}_e(\mathbf{r}, t) \equiv n_e^{-1} \int \mathbf{p} f_e d^3 p , \qquad (2.12)$$

the cold fluid equations for electrons are

$$\partial_t n_e + \nabla \cdot (n_e \mathbf{u}_e) = 0 , \qquad \frac{d\mathbf{p}}{dt} = (\partial_t + \mathbf{u}_e \cdot \nabla)\mathbf{p}_e = -e \left(\mathbf{E} + \frac{\mathbf{u}_e}{c} \times \mathbf{B} \right), (2.13)$$

with $\mathbf{u}_e = \mathbf{p}_e/(m_e \gamma_e c)$ and $\gamma_e = (\mathbf{p}_e^2 + m_e^2 c^2)^{1/2}$. Equations (2.13) are the theoretical basis for the analytic description of the laser-plasma interaction phenomena described in the following. However, in the present paper we do not enter into mathematical details.

2.4 "Relativistic" Optics

2.4.1 Wave Propagation and "Relativistic" Nonlinearities

We consider a transverse EM wave ($\nabla \cdot \mathbf{E} = 0$) propagating in an uniform plasma with electron density n_e. The wave equation for $\mathbf{E}$ is given by

$$\left(\nabla^2 - \frac{1}{c^2} \partial_t^2 \right) \mathbf{E} = \frac{4\pi}{c^2} \partial_t \mathbf{J} , \qquad (2.14)$$

with the current density $\mathbf{J} = -e n_e \mathbf{u}_e$ (ions are assumed as an immobile, neutralizing background). For electron velocities $|\mathbf{u}_e| \ll c$, we pose $\gamma_e \simeq 1$ and neglect the $\mathbf{u}_e \times \mathbf{B}$ term, so that $\mathbf{u}_e$ is proportional to $\mathbf{E}$. This is the basis for the linear optics of a plasma (supposed to be non-magnetized), which can be described by the refractive index $n = n(\omega)$ with

---

contribution of the low-frequency part is negligible with respect to that of the dominant frequencies in the radiation spectrum.

$$\mathsf{n}^2 = \varepsilon = 1 - \frac{\omega_p^2}{\omega^2} = 1 - \frac{n_e}{n_c} \, , \tag{2.15}$$

where $\varepsilon = \varepsilon(\omega)$ is the dielectric function, $\omega_p = (4\pi e^2 n_e/m_e)^{1/2}$ is the plasma frequency and $n_c = m_e \omega^2/4\pi e^2$ is named the cut-off or "critical" density. Wave propagation requires n to be a real number, which occurs when the wave frequency $\omega < \omega_p$ or, equivalently, the plasma density $n_e < n_c$, that defines an *underdense* plasma which is transparent for the frequency ω. If $n_e > n_c$ the plasma is *overdense* and reflecting. For $\lambda_L = 1\,\mu\text{m}, n_c \simeq 10^{21}\,\text{cm}^{-3}$ which falls between the typical densities of gaseous and solid media, respectively.

When the EM wave amplitude is such that $a_0 \gtrsim 1$, nonlinear optical effects arise because of both the dependence of γ_e on the instantaneous field and the importance of the $\mathbf{u}_e \times \mathbf{B}$ term. Thus, the wave propagation depends on its amplitude and higher harmonics of the main frequency are generated.

However, for CP there is a *particular* plane wave, a monochromatic solution for which $\mathbf{u}_e \times \mathbf{B} = 0$ and $\gamma_e = (1 + a_0^2/2)^{1/2}$ is constant in time (this solution is related to the case of the single particle orbits for CP described in Sect. 2.2). In this particular case, the electron equation of motion reduces to

$$\frac{\mathrm{d}\mathbf{p}_e}{\mathrm{d}t} = m_e \gamma_e \frac{\mathrm{d}\mathbf{u}_e}{\mathrm{d}t} = -e\mathbf{E} \, , \tag{2.16}$$

which is identical to the non-relativistic, linearized equation of motion but for the constant factor γ_e that multiplies m_e. Thus we immediately obtain that the wave propagation can be described by the nonlinear refractive index n_{NL} with

$$\mathsf{n}_{\mathrm{NL}}^2(\omega) = 1 - \frac{\omega_p^2}{\gamma_e \omega^2} = 1 - \frac{n_e}{\gamma_e n_c} \, . \tag{2.17}$$

It should be kept in mind that, in general, a nonlinear refractive index should be used with care and that, in particular, (2.17) applies only to the idealized case of a monochromatic CP wave in a homogeneous plasma: already the extension to LP is not straightforward since γ_e is not constant anymore. In the present context, we use (2.17) for a simple description of the phenomenon of "relativistic" *self-focusing*. We also show, however, that applying (2.17) to the other characteristic phenomena of "relativistic" *transparency* leads to incorrect predictions.

2.4.2 Relativistic Self-focusing

We consider a EM beam propagating in a plasma along x. We assume that the beam has a standard bell-shaped profile (e.g., Gaussian), so that the intensity will be highest on the axis and decrease to zero with increasing radial distance. $r_\perp$. Thus, using (2.17) as a function of the local amplitude $\mathbf{a} = \mathbf{a}(x, r_\perp, t)$, i.e. taking $\gamma_e = (1 + \langle \mathbf{a} \rangle^2 /2)^{1/2}$, we obtain that n_{NL} has its *highest* value on the axis ($r_\perp = 0$) and then decreases with increasing radial distance $r_\perp$, down to the linear value (2.15). This implies that the refractive index, due to its nonlinear dependence, is modulated as in an optical fiber or dielectric waveguide, leading to a *self-focusing* (SF) effect which counteracts diffraction.

Figure 2.3a describes a simple SF model based on a geometrical optics description. We assume a "flat top" radial profile so that the intensity is almost constant in the central region. Thus, the refractive index has values $n_a = n_{NL}[\mathbf{a}(r_\perp = 0)]$ for $r_\perp < D/2$, where D is the beam diameter, and $n_b = n_{NL}[\mathbf{a} = 0]$ for $r_\perp > D/2$. Because of diffraction, light rays tend to diverge with a typical angle $\theta_i \simeq \arccos(\lambda/D)$. At the $r_\perp = D/2$ boundary, due to Snell's law the rays are bent to an angle $\theta_r = \arcsin((n_a/n_b)\sin\theta_i)$, with total internal reflection occurring as $\theta_r = \pi/2$. This yields a threshold for the guiding of the beam inside the central region. In the limit of weak nonlinear effects ($|\mathbf{a}| \ll 1$) and small angles ($\lambda/D \ll 1$) the condition can be written as

$$\pi\left(\frac{D}{2}\right)^2 |\mathbf{a}(r_\perp = 0)|^2 \simeq \pi\lambda^2\frac{n_c}{n_e}. \tag{2.18}$$

Note that the first term is proportional to the beam *power*. Inserting numbers and recalling that $\mathbf{a} = e\mathbf{A}/m_e c^2$, one obtains the threshold power value as $P_T \simeq 43\,\mathrm{GW}(n_c/n_e)$. Thus, this rough model predicts the same scaling with density and order of magnitude as the reference value $P_T = 17.5\,\mathrm{GW}(n_c/n_e)$ which is obtained from a more rigorous theory [18]. Notice, however, that also this latter estimate is based on some assumptions, i.e. a CP beam which is several wavelengths wide and long: it may not be applied to ultrashort, tightly focused pulses extending only over a few wavelengths. Also notice that the evolution of a laser pulse undergoing SF may be quite complex; at least, it involves the creation of a low-density channel as

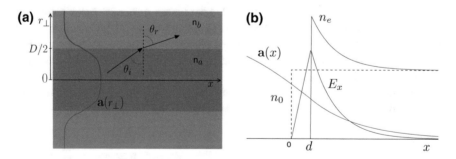

Fig. 2.3 **a** "optical fiber" model of self-focusing. Since the laser beam has a radial intensity profile $\mathbf{a}(r_\perp)$ the nonlinear refractive index has higher values in the central region, causing a guiding effect. **b** evanescence of the EM field $\mathbf{a}(x)$ in an overdense plasma ($n_0 > n_c$) that fills the $x > 0$ region. The electron density (n_e) profile is modified self-consistently by the action of the ponderomotive force which is balanced by the space-charge field E_x

the electrons are pushed away from the axis due to ponderomotive forces (see Sect. 2.2.2).

2.4.3 Relativistic Transparency

Equation (2.17) implies that n_{NL} is real for $n_e > \gamma_e n_c$, i.e. the cut-off density is increased by a factor γ_e with respect to the linear, non-relativistic case. The usual description is that a plasma may become transparent because of relativistic effects, and one often reads of a "relativistically corrected" cut-off density $\gamma_e n_c$.

Indeed, there are two examples of "relativistic" transparency which are of practical importance and where taking $n_e < \gamma_e n_c$ as a criterion for wave propagation leads to erroneous predictions. The first is the case of wave incidence on a semi-infinite plasma with a step boundary. In the linear regime, one may assume the profile of the electron density to be unperturbed, so the problem is reduced to imposing boundary conditions at the plasma-vacuum interface which leads to Fresnel formulas ([9], Sect. 7.3). For strong fields, however, the density profile is modified by the wave action. Taking the simplest case of normal incidence of a CP wave [19], the steady ponderomotive force originating from the cycle-average of the $\mathbf{u}_e \times \mathbf{B}$ term pushes the electrons inside the target and pile them up causing a local increase of the density in the evanescence layer, which counteracts the relativistic effect (Fig. 2.3b). As a consequence, the threshold for wave penetration (for $n_e \gg n_c$ and $a_0 \gg 1$) becomes $a_0 > (\sqrt{3}/2)^3 (n_e/n_c)^2$ [19], which corresponds to much higher intensities than predicted by posing $\gamma_e > n_e/n_c$ i.e. $a_0 > \sqrt{2} n_e/n_c$.

The second example is that of a thin foil of thickness $\ell \ll \lambda = 2\pi c/\omega$, for which the relevant parameter for transparency is the areal density $n_e\ell$. The nonlinear transmission and reflection coefficients can be calculated for a normally incident CP wave by assuming a Dirac delta-like profile [20], showing the onset of transparency when

$$a_0 > \zeta \equiv \pi \frac{n_e}{n_c} \frac{\ell}{\lambda} . \tag{2.19}$$

Thus, for ultrathin targets such that $\ell \ll \lambda$ it is possible to have the onset of transparency even when $n_e > \gamma_e n_c$.

It is worth noticing, however, that also these models are one-dimensional, i.e. based on plane waves. Multi-dimensional effects play an important role for any realistic laser pulse with a finite transverse profile. In particular, the ponderomotive force may reduce the electron density on axis by pushing electrons away, enhancing the penetration of the laser pulse.

2.5   Interaction With a Step Boundary Plasma

We now focus on the interaction of a superintense laser pulse with a strongly over-dense plasma ($n_e \gg n_c$) having a step-like density profile, e.g. $n_e \simeq n_0 \Theta(x)$ with $\Theta(x)$ the Heaviside step function. This problem is relevant to experiments on the interaction of ultrashort pulses with solid targets.

2.5.1   Energy Absorption: From Fresnel Formulas to "Vacuum Heating"

In the linear regime, the solution for the problem of the interaction between a plane EM wave and a medium having refractive index n and a steep interface is provided by the matching relations for the wavevectors and by Fresnel formulas for the reflection and absorption coefficients, which depend on the angle of incidence and the wave polarization. Using (2.15) for n one finds that inside the medium ($x > 0$, for definiteness) the wave is evanescent as e^{-x/ℓ_s} with $\ell_s = c(\omega_p^2 - \omega^2)^{-1/2}$ and there is total reflection of the incident energy since n is purely imaginary, which corresponds to neglecting any dissipative process. Dissipation may be provided by resistivity due to Coulomb collisions between electron and ions (Drude model), so that (2.15) is modified by replacing $\omega^2 \to \omega(\omega + i\nu_{ei})$ where ν_{ei} is the collision frequency. However, ν_{ei} quickly decreases with increasing electron energy ("runaway effect") making collisional absorption inefficient at high intensities.

Actually, there are *collisionless* mechanisms taking place in the surface region of evanescent field (the "skin layer") which may produce a sizable absorption (see e.g. [21] and references therein). The essence of such mechanisms is that in crossing the skin layer an electron sees the evanescent field to change in a time shorter than the oscillation period $2\pi/\omega$, so that $\langle \mathbf{v}(t) \cdot \mathbf{E}(x = x(t), t) \rangle \neq 0$ over the electron trajectory $x(t)$. Calculating the total absorption requires a kinetic approach. However, to some extent, collisionless skin layer absorption might be included phenomenologically in the Fresnel modeling by replacing ν_{ei} with an effective collision frequency.

Indeed, at very high intensities absorption may be due to the generation of energetic electrons through a mechanism which violates a basic underlying assumption of the Fresnel modeling, i.e. that all electrons remain into the $x > 0$ region initially occupied by the plasma. Depending on the EM wave polarization, there can be an oscillating Lorentz force component perpendicular to the surface, so that for strong enough fields an electron can be driven from the plasma surface into the vacuum region (Fig. 2.4). After half a period of the driving force, the electron re-enters into the plasma region with a finite velocity and may cross the evanescence layer, thus escaping from the accelerating field region and being "absorbed" in the plasma. During the half-oscillation on the vacuum side, the electron acquires an energy of the

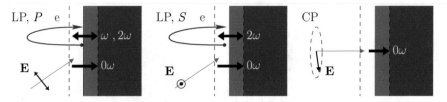

Fig. 2.4 Oscillatory and steady forces on an overdense plasma with steep boundary, for different polarizations. For linear polarization (LP) with **E** in the plane of incidence (*P*-polarization), both the **E** and **v** × **B** terms in the Lorentz force can drive electron "half-oscillations" across the plasma-vacuum interface at a rate ω and 2ω, respectively. For **E** perpendicular to the plane of incidence (*S*-polarization) only the **v** × **B** term drives the half-oscillations. For circular polarization (CP) and normal incidence, all the oscillating force components perpendicular to the surface are suppressed. In all cases, there is a steady ("0ω") force pushing the electrons and giving rise to radiation pressure action on the plasma

order of the oscillation energy in the wave field,[5] i.e. $\mathcal{E}_e \simeq m_e c^2 \left((1 + \langle \mathbf{a}^2 \rangle)^{1/2} - 1 \right)$. This is the essential description of the mechanism originally proposed by Brunel [22] and widely referred to as "vacuum heating" (VH). Brunel originally considered the electric field component for *P*-polarization as the driver for electron half-oscillations across the surface, so that energetic electron bunches are generated once per laser cycle. A simple model [2] yields for the reflectivity R the following implicit relation

$$R \simeq 1 - \frac{1 + \sqrt{R}}{\pi a_0} \left(\left(1 + (1 + \sqrt{R})^2 a_0^2 \sin^2 \theta_i \right)^{1/2} - 1 \right) \frac{\sin \theta_i}{\cos \theta_i} , \qquad (2.20)$$

with θ_i the incidence angle. In the $a_0 \sin \theta_i \ll 1$ limit, $R \simeq 1 - (4/\pi) a_0 \sin^3 \theta_i / \cos \theta_i$.

The magnetic component of the Lorentz force can also act as driver, so that VH may take place also for *S*-polarization and normal incidence generating electron bunches twice per laser cycle (since the magnetic force term has frequency 2ω). This is also referred to as "**J** × **B**" heating, although the name comes from an earlier suggestion about the contribution of the magnetic force to absorption [23]. Instead, for *circular* polarization and normal incidence there is no oscillating component normal to the surface[6] so that electron heating may be suppressed [24].

2.5.2 Momentum Absorption and Radiation Pressure

In addition to energy, EM field contain traslational momentum, its density being $\mathbf{g} = \mathbf{E} \times \mathbf{B}/4\pi c$. Thus, an idealized quasi-plane-wave "square" pulse of duration τ and

---

[5]This estimate for the electron energy is commonly referred to as "ponderomotive scaling"; probably, the name originates from the questionable definition of nonlinear oscillating forces as "ponderomotive" (Sect. 2.2.2).

[6]This is analogous to the absence of high-frequency longitudinal motion in a CP wave, Sect. 2.2.1.

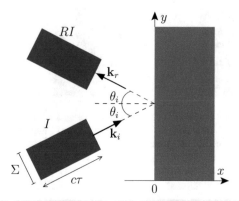

Fig. 2.5 Simple kinematic model to calculate the EM momentum transfer through a reflecting surface and the resulting radiation pressure. A "box-shaped", quasi-plane wave pulse of intensity I, duration τ and transverse section Σ impinges at an angle θ_i on the surface. If the latter is at rest, the reflected pulse is in the specular direction, has the same duration and section as the incident pulse, and an intensity RI where R is the reflectivity of the surface

transverse area Σ (Fig. 2.5) contains a total momentum $\mathbf{p}_i = \mathbf{g}\Sigma c\tau = (I/c^2)(\Sigma c\tau)\hat{\mathbf{n}}$ where $I = (c/4\pi)|\mathbf{E} \times \mathbf{B}|$ is the intensity and $\hat{\mathbf{n}}$ the direction of propagation. Under reflection from the surface of a medium with reflectivity R, momentum is transferred to the medium giving rise to a net force perpendicular to the surface, i.e. to radiation pressure. By simple kinematic relations, the pressure on the surface can be obtained as

$$P_\perp = (1 + R)\frac{I}{c}\cos^2\theta_i \ , \tag{2.21}$$

where we took $\hat{\mathbf{n}} = (\cos\theta_i, \sin\theta_i)$ and the surface at $x = 0$. The maximum pressure of $2I/c$ is obtained for a perfect mirror ($R = 1$) at normal incidence ($\theta_i = 0$). The above relations are of classical nature, however one may also obtain the radiation pressure kinematically by describing the incident pulse as a bunch of N photons each of energy $\hbar\omega$ and momentum $(\hbar\omega/c)\hat{\mathbf{n}}$ of which a fraction R is elastically reflected at the surface. The classical expression is recovered by the equation for the pulse/bunch energy $I\Sigma c\tau = N\hbar\omega$.

Going back to the classical description, one can also obtain the total pressure from the knowledge of the EM fields by integrating the total force per unit volume over the whole plasma,

$$P_\perp = \int_0^{+\infty} \left(\rho\mathbf{E} + \frac{\mathbf{J}}{c} \times \mathbf{B}\right) \cdot \hat{\mathbf{x}}\mathrm{d}x \ , \tag{2.22}$$

where ρ is the charge density. To test a simple case, we may assume normal incidence ($\theta_i = 0$) so that $\mathbf{E} \cdot \hat{\mathbf{x}} = 0$, and calculate the fields inside the plasma in the linear limit by using Fresnel formulas with n given by (2.15) so that $R = 1$. In this case, besides recovering easily the result $P_\perp = 2I/c$ one observes that the integrand of (2.22) is the non-relativistic ponderomotive force (2.5) multiplied by n_e. In practice the local ponderomotive force is on the electrons only (the $\mathbf{v} \times \mathbf{B}$ term on ions is smaller by a factor $\sim m_e/m_i \sim 10^{-3}$), but as soon as the force pushes the electrons in the region of evanescent fields, a charge depletion layer is created at the surface with an electrostatic field which back-holds electrons and exerts a force on ions in the inward direction. This situation is evidenced in Fig. 2.3b which shows the charge separation layer $(0 < x < d)$ and the corresponding electrostatic field E_x. If the electrons are in equilibrium, the ponderomotive force is exactly balanced locally by the electrostatic one, so in turn the ions feel an electrostatic pressure which equals the radiation pressure value. In the absence of counteracting forces, the electrostatic field will accelerate ions, so that ultimately the EM momentum is transferred to the whole medium. Radiation pressure of superintense lasers is currently investigated as a driving mechanism for laser-plasma accelerators of ions [7]: related concepts are investigated in Sect. 2.6.

2.5.3 Absorption of Tangential Momentum

By applying the same kinematics leading to (2.22), we also obtain that for a medium with partial reflectivity $(R < 1)$ there is absorption of EM momentum also in the *parallel* direction, i.e. along the surface, yielding a *tangential* pressure.

$$P_{\parallel} = (1 - R)\frac{I}{c} \sin \theta_i \cos \theta_i . \tag{2.23}$$

We thus expect that (referring to the two-dimensional, plane wave geometry of Fig. 2.5) the ponderomotive force has a tangential (y) component F_{py}, which can drive a surface current j_y of electrons. Such surface current has been often observed in simulations since early studies of absorption at oblique incidence [25] but, to our knowledge, no simple model was presented until recently; below we resume the basic findings of our model [26] which were partly anticipated in [27].

If the plasma is homogeneous along y, the current j_y produces no charge separation and thus no electrostatic field. Indeed, j_y generates a magnetic field B_z which, while growing in time, induces an electric field E_y which counteracts the ponderomotive action. However, the evanescence lengths of F_{py} and E_y are different, so that the ponderomotive and electric forces cannot balance locally and a double layer of current is generated, which leads to a B_z localized in the skin layer. For an incident EM wave with flat-top profile, i.e. having constant intensity $I = I_0$ for $0 \le t < \tau_L$, both j_y and B_z are found to grow linearly in time until $t = \tau_L$ with the maximum value of B_z at the time t being

$$B_z^{(\text{max})} \simeq \frac{\pi}{6}\frac{t}{\tau_L}(1 - R)\sin(2\theta_i)a_0 B_L , \tag{2.24}$$

where $a_0 = I_0/m_e n_c c^3$ and B_L are the dimensionless and magnetic field amplitudes, respectively, of the incident wave. Intense laser pulses ($a_0 \gg 1$) can yield high absorption and low reflectivities down to $R \simeq 0.5$, so that the amplitude of the slowly-varying field B_z may approach that of the laser field B_L, i.e. $\simeq 10^9$ Gauss for $a_0 \sim 10$.

2.6 Moving Mirrors

The picture of "vacuum heating" presented in Sect. 2.5.1, in which electrons are periodically dragged out of and back into the plasma, is oversimplified. In reality the oscillating components of the Lorentz force drive a collective oscillation of the electron density profile (with the high energy electron bunches being related to the partial "breaking" of such oscillations). We may thus assume that the $n_e = n_c$ surface oscillates back and forth under the action of the Lorentz force. Thus, the incident laser pulse is reflected from a surface whose position oscillates either at the same frequency of the laser, or twice that value depending on the incidence angle and polarization. If we consider instead the action of the time-averaged force, i.e. of radiation pressure, the $n_e = n_c$ surface is pushed inwards, so we have reflection from a surface moving along the propagation direction. The relativistic *moving mirror* model is able to explain (at least qualitatively) basic features of both the above mentioned scenarios, which are relevant to important applications of superintense interaction with overdense plasmas (e.g. solid targets). It is thus worth to review here some basic relations of reflection from a moving mirror.

2.6.1 Reflection from a Moving Mirror

For brevity and simplicity we consider normal incidence only and we assume a "perfect" mirror whose reflectivity $R = 1$ in its *rest frame*. Let the mirror move with velocity $\mathbf{V} = V\hat{\mathbf{x}}$ and an EM plane wave of frequency ω, field amplitude E_i and intensity $I = (c/4\pi)E_i^2$ be incident from the $x < X_m$ side, where X_m is the mirror position (Fig. 2.6). For the moment we assume V to be constant, hence $X_m = Vt$.

The laws of reflection are known in the rest frame of the mirror (L'): the EM wave is reflected with inversion of both the wavevector and the electric field and no change of frequency. Thus we can obtain the frequency ω_r and the amplitude E_r of the reflected wave in the lab frame (L) by a first Lorentz transformation of the incident wave from L to L', and then by a second transformation of the reflected wave from L' to L. The result is

$$\frac{\omega_r}{\omega} = -\frac{E_r}{E_i} = \frac{1-\beta}{1+\beta}, \tag{2.25}$$

where $\beta = V/c$. Thus, if $V > 0$, i.e. if the EM wave propagates in the same direction as the mirror velocity, the frequency is "red-shifted" towards lower values and the amplitude is also lower than for the incident pulse. If $V < 0$, i.e. if the wave is counterpropagating with respect to the mirror, "blue-shift" and amplitude increase occur. In the highly relativistic limit ($\beta \to 1$) notice that $(1 - \beta)/(1 + \beta) \simeq (2\gamma)^{-2}$.

The above relations might also be found by noticing that, for normal incidence (and thus the electric field parallel to the mirror surface) the boundary condition $\mathbf{E}'(x' = X'_m)$ for a perfect mirror at rest in L' corresponds to $\mathbf{A}(x = X_m) = 0$ in L for arbitrary motion $X_m = X_m(t)$, as can be easily demonstrated via a Lorentz transformation and the relations between $\mathbf{A}$, $\mathbf{E}$ and $\mathbf{B}$. Thus, by posing

$$\left[A_i e^{ikx - i\omega t} + A_r e^{-ik_r x - i\omega_r t} \right]_{x = Vt} = 0, \tag{2.26}$$

where $k = \omega/c$ and $k_r = \omega_r/c$, (2.25) are obtained again.

If we consider an incident pulse of long but finite duration τ, such as the "square" packet in Fig. 2.6, the number of oscillations inside the pulse is a Lorentz invariant. Thus, the duration of the reflected pulse is $\tau_r = \tau(1 + \beta)/(1 - \beta)$, i.e. $\tau_r > \tau$ if $V > 0$ and $\tau_r < \tau$ if $V < 0$. Since the intensity of the reflected field is $I_r = I(1 - \beta)^2/(1 + \beta)^2$, we find that $I_r \tau_r < I\tau$ for $V > 0$, i.e. the incident pulse loses energy to the mirror, while the opposite occurs for $V < 0$. A counterpropagating mirror may thus be used to both compress in time and amplify an incident pulse: an intriguing laser-plasma based scheme of such kind has been proposed as a way to reach unprecedentedly high intensities [28].

2.6.2   High Harmonics from an Oscillating Mirror

Now suppose the perfect mirror performs an oscillatory motion, $X_m = X_0 \sin \Omega t$. To find the reflected field we can use again the condition $A(x = X_m, t) = 0$ and thus

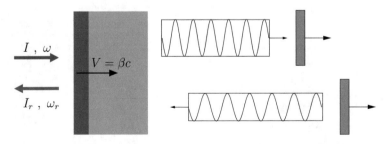

Fig. 2.6 EM wave of intensity I and frequency ω impinging on a moving mirror. When the mirror velocity $\mathbf{V}$ is in the propagation direction as in the picture, the wave frequency is red-shifted and a reflected pulse has longer duration and lower energy than the incident pulse. Conversely, blue-shift and energy increase occur for a counter-propagating mirror

write, e.g.,

$$0 = [A_i(x,t) + A_r(x,t)]_{x=X_0 \sin \Omega t} = [A_i \cos(kx - \omega t) + A_r(x,t)]_{x=X_0 \sin \Omega t} \,, \tag{2.27}$$

from which we obtain, using some math, that the temporal dependence of the reflected pulse is

$$A_r(t) \sim \sin\left(\omega t + \frac{2\omega}{c} X_0 \sin \Omega t\right) \sim \sum_{n=0}^{\infty} J_n\left(\frac{2\omega X_0}{c}\right) \sin(\omega + n\Omega)t \,, \tag{2.28}$$

where the J_n's are Bessel functions. Thus, the reflected wave contains a mixing of ω, the frequency of the incident wave, with integer harmonics of the mirror frequency, Ω.

An intense laser pulse of frequency ω drives oscillations of the surface of an overdense plasma at frequency ω or 2ω depending on the angle of incidence and the polarization (Fig. 2.7b). The moving mirror model thus predicts that a P-polarized pulse will generate P-polarized harmonics at all integer frequencies of the driving pulse (ω, 2ω, 3ω, ...) while a S-polarized pulse will generate only odd frequencies $(2n + 1)\omega$. Of course, since the mirror is driven by the same laser pulse it reflects, any

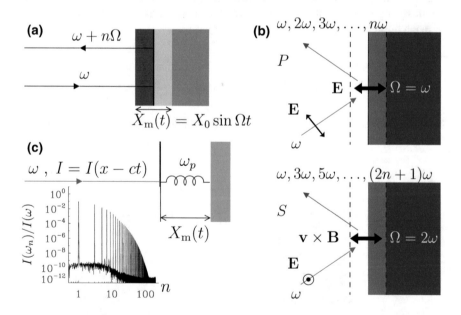

Fig. 2.7 Oscillating mirrors and harmonic generation. **a** frequency mixing in the reflected wave. **b** driving of a plasma surface at different frequencies depending on the polarization and incidence angle, leading to the generation of different order of harmonics. **c** a toy model for a laser-driven oscillating mirror. The inset shows a spectrum of the reflected pulse obtained with such a model (n is the harmonic order)

estimate of the intensity of such harmonics must be based on some self-consistent modeling for dynamics of the moving mirror. A toy model might be formulated by assuming that the mirror is bound by a spring of frequency ω_p (Fig. 2.7c), which roughly accounts for the resonant plasma response, and by inserting a friction term to phenomenologically account for finite absorption. For a mirror driven by a linearly polarized, "flat-top" (constant intensity I) pulse at normal incidence, the equation of motion is

$$\frac{d}{dt}(\gamma_m \beta_m) = \frac{2I}{\sigma c^2}(1 + 2\cos(2\omega t_r))\frac{1 - \beta_m}{1 + \beta_m} - \omega_p^2 X_m - \nu_m \beta_m c, \qquad (2.29)$$

where $dX_m/dt = \beta_m c$ and $t_r = t - X_m/c$. In (2.29) σ is the mass per unit area of the mirror, so that when referring to an oscillating plasma surface we might roughly estimate $\sigma \simeq m_e n_e \ell_s$ with ℓ_s the evanescence length (ions are assumed to be at rest). Equation (2.29) may be easily solved numerically to obtain the maximum velocity of the mirror $\beta_{max}c$, which according to (2.25) should be related to the spectral cut-off frequency $\omega_{co} \simeq 4\omega \gamma_{max}^2$ when $\beta_{max} \to 1$. Thus, if $\gamma \sim (1 + a_0^2)^{1/2}$ one expects to generate harmonics up to orders $\sim 10^2$ with state-of-the-art lasers. One can also obtain, via (2.26), the temporal profile of the reflected pulse. The latter usually appears as a train of ultrashort spikes, which can be qualitatively understood as a coherent modulation of the incident pulse waveform by the moving mirror: each semicycle is alternatively stretched or compressed depending on the sign of $\beta_m(t)$. A quantitative description of high harmonic generation needs a more realistic modeling and simulations of the laser-plasma dynamics, of course (see [29, 30] for reviews).

2.6.3 Light Sail Acceleration

Now assume a thin plane mirror of mass density ρ_m and thickness ℓ, and a plane wave pulse $I = I(t)$ at normal incidence and with circular polarization so that there are no oscillating components. The mirror is thus accelerated by radiation pressure according to the equation of motion

$$\frac{d}{dt}(\gamma_m \beta_m) = \frac{2I(t_r)}{\rho \ell c^2}R(\omega')\frac{1 - \beta_m}{1 + \beta_m}, \qquad (2.30)$$

which we name the *light sail* (LS) equation. As we consider the acceleration of the foil as a whole,[7] with respect to (2.29) there are no elastic and friction terms. Instead, we include a finite reflectivity $R < 1$ to account for partial transmission through the foil. Notice that in general R depends on the incident pulse frequency and it is defined for a mirror at rest, thus it is a function of the frequency in the moving frame $\omega' = \omega(1 - \beta_m)^{1/2}(1 + \beta_m)^{-1/2}$ and, for a thin ($\ell \ll \lambda$) plasma mirror it is

---

[7]Note that $\rho_m \ell$ in (2.30) is formally equivalent to σ in (2.29), but here in (2.30) $\rho_m \ell$ refers to the total mass of the mirror, i.e. including the ions.

proportional to $\rho\ell$. At intensities high enough for relativistic transparency effects to be important, R quickly drops from unity as the threshold in (2.19) is exceeded, so that $a_0 \simeq \zeta$ is an optimal compromise between reducing the areal mass and increasing reflectivity at fixed thrust in order to maximize the sail acceleration. In the following we assume for simplicity $R = 1$ although an analytic solution of (2.30) may be found also for a partially transparent "delta-like" foil [31].

From (2.30) the final γ-factor is obtained as

$$\gamma_m(t = \infty) - 1 = \frac{\mathcal{F}^2}{(2(\mathcal{F}+1))} , \qquad \mathcal{F} = \frac{2}{\rho\ell} \int_0^\infty I(t')dt' , \tag{2.31}$$

where $\mathcal{F}$ can be estimated as a function of the average intensity I and pulse duration τ,

$$\mathcal{F} = \frac{2I\tau}{\rho\ell} = \frac{Z}{A} \frac{m_e}{m_p} \frac{a_0^2}{\zeta} \omega\tau . \tag{2.32}$$

We thus see that present-day femtosecond lasers having $\tau \sim 10(2\pi/\omega)$ and $a_0 \sim 10$ are in principle able to accelerate ultrathin targets up to $\gamma_m - 1 \gtrsim 0.1$, which corresponds to an energy per nucleon exceeding 100 MeV, while future lasers yielding $a_0 \sim 10^2$ could drive relativistic GeV nuclei. In addition, LS acceleration becomes more efficient with increasing speed, the mechanical efficiency η_{mec} (ratio of sail energy $\mathcal{E}_{\text{LS}}$ over driver pulse energy $I\tau$, all defined per unit surface) being

$$\eta_{\text{mec}} \equiv \frac{\mathcal{E}_{\text{LS}}}{I\tau} = \frac{2\beta_m}{1 + \beta_m} . \tag{2.33}$$

This relation can be obtained from (2.30), but also from a simple quantum picture taking the pulse as a bunch of $\mathcal{N}$ photons (per unit surface) whose energy drops from $\hbar\omega$ to $\hbar\omega_r$ due to reflection from the sail. Thus, since $\mathcal{N} = I\tau/\hbar\omega$,

$$\mathcal{E}_{\text{LS}} = \mathcal{N}\hbar(\omega - \omega_r) = \mathcal{N}\hbar\omega \frac{2\beta_m}{1 + \beta_m} = \eta_{\text{mec}} I\tau . \tag{2.34}$$

The efficiency of LS acceleration is what makes it attractive for interstellar propulsion of probes from Earth [32] as well for laser-driven ion accelerators [7]. For this latter application, additional features as monoenergetic spectrum and ultrashort duration (since ideally all ions in the sail propagate at the same velocity) make the LS appear as a "dream bunch" of energetic ions. Issues include the slow energy gain, since (2.30) shows that the force on the sail decreases with increasing β_m so that reaching the highest possible energy requires stability over long distances. The modeling in a realistic geometry brings both good news (LS might be faster and more efficient in 3D than in 1D [33, 34], which is uncommon) and bad news (the sail might be prone to Rayleigh-Taylor-type instabilities [35, 36], see Sect. 2.7).

2.7 Instabilities

Instability is maybe the word which is more frequently associated to *plasma*, the obvious reason being that the main obstacle to achieving controlled fusion is that a plasma tends to become unstable in several ways, quickly destroying the desired configuration. The basic laser-plasma interaction processes we reviewed so far (as well as other we did not include) may also lead to, or be affected by instabilities. For example, a laser pulse greatly exceeding the power threshold for relativistic self-focusing may break up in multiple filaments, especially if its intensity distribution is not smooth. As another example, the high-energy electrons produced by laser-plasma interactions typically lead to an anisotropical distribution function which is unstable against electromagnetic perturbations (Weibel instability): the growth of the latter act to deviate particle trajectories in order to create a more isotropic distribution. In the context of laser-plasma interactions one also encounters nonlinear processes where a strong "pump" mode having frequency ω_0 and wavevector $\mathbf{k}_0$, such as e.g. an intense laser pulse propagating in the plasma or an high amplitude plasma wave, excites two (or more) "daughter" plasma modes whose frequencies and wavevectors are related by the phase matching relations $\omega_0 = \omega_1 + \omega_2$ and $\mathbf{k}_0 = \mathbf{k}_1 + \mathbf{k}_2$. These processes are referred to as *parametric instabilities* since the daughter modes may also grow at high amplitude at a rate typically proportional to the amplitude of the pump mode. An example is Raman backscattering with corresponds to a laser wave exciting a plasma wave and an EM wave in the backward direction, which can lead to strong reflection from a low density plasma.

Covering all the possible instabilities in the laser-plasma scenario is much beyond the limits and scope of the present paper, thus we just give some further detail on instabilities affecting the dynamics of the moving mirror dynamics outlined in Sect. 2.6. The plasma surface oscillating under the action of the Lorentz force has been found in simulations to develop ripples which also oscillate at half the driving frequency [37]. This is due to a parametric instability in which the driven surface oscillation decays into two surface waves, similarly to the phenomenon of Faraday ripples (or waves)[8] originating on the surface of a fluid subject to vertical vibrations. In the context of laser-plasma interaction the effect was studied in relation to the onset of surface rippling in experiments on high harmonic generation, where the harmonic emission was observed to turn from collimated to diffuse over a certain intensity threshold.

When the plasma surface is steadily accelerated by radiation pressure as in the light sail concept (Sect. 2.6.3), rippling may occur because of an instability of the Rayleigh-Taylor (RT) type. The simplest example of RT instability (RTI) is that of an heavy fluid of density ρ_2 placed above a lighter one of density $\rho_1 < \rho_2$ in a gravity field $\mathbf{g}$ (Fig. 2.8): a small perturbation at the surface lowers the energy of the system and thus grows up exponentially ($\sim e^{\gamma_{RT} t}$) in a first stage, favoring the mixing of the two fluids. The equivalence principle tells us that the same effect is produced in the presence of an acceleration field $\mathbf{a}$ directed from the light fluid to the heavier one:

---

[8]https://en.wikipedia.org/wiki/Faraday_wave.

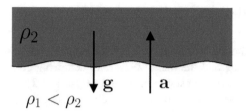

Fig. 2.8 Rayleigh-Taylor instability: an interface between two fluids of different mass density becomes corrugated in the presence of a gravity field anti-parallel to the density gradient or, equivalently, an acceleration parallel to the density gradient

this is the instability form which strongly affects the compression of fuel pellet in Inertial Confinement Fusion [38].

For a sinusoidal perturbation of wavevector k_{RT}, the RTI growth rate is given by (see e.g. [39])

$$\gamma_{RT} = \left(ak_{RT}\frac{\rho_2 - \rho_1}{\rho_2 + \rho_1}\right)^{1/2}, \tag{2.35}$$

where $a = |\mathbf{g}|$ in the case of the gravitational RTI. The case of a plasma surface accelerated by radiation pressure can be viewed as a massless fluid of photons pushing a heavy material fluid, and it is thus unstable with a rate $\gamma_{RT} = (ak_{RT})^{1/2}$. RTI also occurs for a thin interface layer separating two fluids of different pressures, which matches closely the LS scenario where the target is placed between the photon fluid and vacuum. The growth rate of such RTI, for non-relativistic dynamics, has the same form as the preceding formula with $a = (2I/\rho\ell c)$ [40]. Analytical models accounting for relativistic motion and other effects can be found, e.g., in [41, 42]. These works left open the question why the surface rippling often observed in simulations occurs predominantly for a wavevector $k_{RT} \simeq 2\pi/\lambda$, i.e. with a periodicity close to the laser wavelength. In [35, 36] it has been suggested that the rippling of the surface self-modulates the radiation pressure, so that depending on the laser polarization the accelerating force may become stronger in the valleys of the ripples and boost their growth. The effect is strongest for a sinusoidal rippling at the laser wavelength because of a resonant coupling with surface plasma waves.

2.8 Angular Momentum Absorption and Magnetic Field Generation

The fact that an EM wave carries energy and momentum becomes very eye-catching for superintense laser pulses which, as we saw in the preceding section, can heat matter to extremely high temperatures and accelerate a quite macroscopic object to velocities approaching the speed of light. An EM wave with CP also carries angular momentum which, when absorbed by a sample of matter, may cause its rotation. For a CP laser beam of frequency ω, propagating along x and having a radial intensity profile $I(r)$, the density of angular momentum along the x-direction is

$$\mathcal{L}_x = (\mathbf{r} \times \mathbf{g})_x = -\frac{r}{2c\omega}\partial_r I(r) , \qquad (2.36)$$

where $\mathbf{g}$ is the density of traslational momentum (Sect. 2.5.2). Notice that for a standard bell-shaped profile $\mathcal{L}_x$ peaks at the edge of the beam. The total angular momentum L_x is proportional to the power P of the beam,

$$L_x = \int_0^\infty \mathcal{L}_x(r)2\pi r dr = \frac{1}{c\omega}\int_0^\infty I(r)2\pi r dr = \frac{P}{c\omega} . \qquad (2.37)$$

We have seen in Sect. 2.6 than in the reflection from a perfect mirror an EM wave delivers twice of its traslational momentum, and that if the mirror moves at relativistic velocities most of the EM wave energy is converted into mechanical energy of the mirror. However, it can be shown that *no* angular momentum is transferred to the mirror. The reasoning is very simple by taking a quantum point of view: the value of the "spin" angular momentum of a photon is $\hbar$, independently of the frequency, and in the reflection the spin is not reversed while the number of photons is conserved for a perfect mirror, so there is no net absorption of angular momentum.

In general, absorption of EM angular momentum requires a dissipative mechanism which "destroys" part of the incident photons. At moderate intensities such mechanism is provided by collisions [43]. At extremely high intensities, strong losses by incoherent emission of radiation imply the absorption of many laser photons for each high frequency photon emitted, hence the transfer of angular momentum might become very efficient in a regime dominated by radiation friction effects [44].

The angular momentum of a laser beam is directly absorbed by the electrons, and the associated torque drives an azimuthal electron current. In turn, this current generates an axial magnetic field: this is known as the *inverse Faraday effect* (IFE) even if this is somewhat a misnomer. Even with a steady absorption, the axial field cannot grow indefinitely since it is accompanied by the induction of a solenoidal electric field that counteracts the electron rotation and exerts a torque on ions, which ultimately absorb most of the angular momentum. The mechanism is thus similar to that leading to the absorption of transverse momentum (Sect. 2.5.3). The scaling of the peak magnetic field on axis B_{ax} with laser and plasma parameters is found to be [43, 44]

$$B_{ax} \sim \eta\frac{n_c}{n_e}\frac{c\tau\lambda^2}{D^2 L}B_0 a_0^2 , \qquad (2.38)$$

where η is the absorbed fraction of the laser energy, L is the length over which absorption occurs, $B_0 = m_e c\omega/e$ and other parameters are as previously defined. Notice that $B_0 a_0 = B_L$, the magnetic field amplitude of the laser pulse. Simulations with radiation friction included [44] of the interaction of superintense pulses with overdense plasmas have shown strong radiation losses with η up to 25% and a scaling

$\eta \sim a_0^3$, so that $B_{\mathrm{ax}} \sim a_0^4$. In the simulated conditions, which might be accessible with next-generation lasers, the generation via IFE of magnetic fields of several 10^9 Gauss is observed, providing in the meantime a demonstration of a macroscopic effect of radiation friction.

References

1. A. Macchi, *A Superintense Laser-Plasma Interaction Theory Primer*. Springer Briefs in Physics (Springer, 2013). https://doi.org/10.1007/978-94-007-6125-4
2. P. Gibbon, *Short Pulse Laser Interaction with Matter* (Imperial College Press, 2005)
3. P. Mulser, D. Bauer, *High Power Laser-Matter Interaction*. Springer Tracts in Modern Physics (Springer, 2010). https://doi.org/10.1007/978-3-540-46065-7
4. G.A. Mourou, T. Tajima, S.V. Bulanov, Rev. Mod. Phys. **78**, 309 (2006). https://doi.org/10.1103/RevModPhys.78.309
5. P. Gibbon, Rivista del Nuovo Cimento **35**, 607 (2012). https://doi.org/10.1393/ncr/i2012-10083-8
6. E. Esarey, C.B. Schroeder, W.P. Leemans, Rev. Mod. Phys. **81**, 1229 (2009). https://doi.org/10.1103/RevModPhys.81.1229
7. A. Macchi, M. Borghesi, M. Passoni, Rev. Mod. Phys. **85**, 751 (2013). https://doi.org/10.1103/RevModPhys.85.751
8. S.C. Wilks, W.L. Kruer, M. Tabak, A.B. Langdon, Phys. Rev. Lett. **69**, 1383 (1992). https://doi.org/10.1103/PhysRevLett.69.1383
9. J.D. Jackson, *Classical Electrodynamics* (Wiley, New York, 1998)
10. L.D. Landau, E.M. Lifshitz, *The Classical Theory of Fields*, 2nd edn., chap. 76 (Elsevier, Oxford, 1975)
11. M. Tamburini, F. Pegoraro, A.D. Piazza, C.H. Keitel, A. Macchi, New J. Phys. **12**, 123005 (2010). https://doi.org/10.1088/1367-2630/12/12/123005
12. A. Di Piazza, Lett. Math. Phys. **83**, 305 (2008). https://doi.org/10.1007/s11005-008-0228-9
13. A. Macchi, Physics **11**, 13 (2018). https://physics.aps.org/articles/v11/13
14. C.K. Birdsall, A.B. Langdon, *Plasma Physics Via Computer Simulation* (Institute of Physics, Bristol, 1991)
15. T.D. Arber, K. Bennett, C.S. Brady, A. Lawrence-Douglas, M.G. Ramsay, N.J. Sircombe, P. Gillies, R.G. Evans, H. Schmitz, A.R. Bell, C.P. Ridgers, Plasma Phys. Contr. Fus. **57**, 113001 (2015). https://doi.org/10.1088/0741-3335/57/11/113001
16. J. Derouillat, A. Beck, F. Prez, T. Vinci, M. Chiaramello, A. Grassi, M. Fl, G. Bouchard, I. Plotnikov, N. Aunai, J. Dargent, C. Riconda, M. Grech, Computer Phys. Comm. **222**, 351 (2018). https://doi.org/10.1016/j.cpc.2017.09.024
17. M. Vranic, J. Martins, R. Fonseca, L. Silva, Comp. Phys. Comm. **204**, 141 (2016). https://doi.org/10.1016/j.cpc.2016.04.002
18. G.Z. Sun, E. Ott, Y.C. Lee, P. Guzdar, Phys. Fluids **30**, 526 (1987). https://doi.org/10.1063/1.866349
19. F. Cattani, A. Kim, D. Anderson, M. Lisak, Phys. Rev. E **62**, 1234 (2000). https://doi.org/10.1103/PhysRevE.62.1234
20. V.A. Vshivkov, N.M. Naumova, F. Pegoraro, S.V. Bulanov, Phys. Plasmas **5**, 2727 (1998). https://doi.org/10.1063/1.872961
21. W. Rozmus, V.T. Tikhonchuk, R. Cauble, Phys. Plasmas **3**, 360 (1996). https://doi.org/10.1063/1.871861
22. F. Brunel, Phys. Rev. Lett. **59**, 52 (1987). https://doi.org/10.1103/PhysRevLett.59.52
23. W.L. Kruer, K. Estabrook, Phys. Fluids **28**, 430 (1985). https://doi.org/10.1063/1.865171
24. A. Macchi, F. Cattani, T.V. Liseykina, F. Cornolti, Phys. Rev. Lett. **94**, 165003 (2005). https://doi.org/10.1103/PhysRevLett.94.165003

25. F. Brunel, Phys. Fluids **31**, 2714 (1988). https://doi.org/10.1063/1.867001
26. A. Macchi, A. Grassi, F. Amiranoff, C. Riconda, arXiv e-prints arXiv:1903.10393 (2019)
27. A. Grassi, M. Grech, F. Amiranoff, A. Macchi, C. Riconda, Phys. Rev. E **96**, 033204 (2017). https://doi.org/10.1103/PhysRevE.96.033204
28. S.V. Bulanov, T. Esirkepov, T. Tajima, Phys. Rev. Lett. **91**, 085001 (2003). https://doi.org/10.1103/PhysRevLett.91.085001
29. U. Teubner, P. Gibbon, Rev. Mod. Phys. **81**, 445 (2009). https://doi.org/10.1103/RevModPhys.81.445
30. C. Thaury, F. Quéré, J. Phys. B At. Mol. Opt. Phys. **43**, 213001 (2010). https://doi.org/10.1088/0953-4075/43/21/213001
31. A. Macchi, S. Veghini, T.V. Liseykina, F. Pegoraro, New J. Phys. **12**, 045013 (2010). https://doi.org/10.1088/1367-2630/12/4/045013
32. Z. Merali, Science **352**(6289), 1040 (2016). https://doi.org/10.1126/science.352.6289.1040
33. S.V. Bulanov, E.Y. Echkina, T.Z. Esirkepov, I.N. Inovenkov, M. Kando, F. Pegoraro, G. Korn, Phys. Rev. Lett. **104**, 135003 (2010). https://doi.org/10.1103/PhysRevLett.104.135003
34. A. Sgattoni, S. Sinigardi, A. Macchi, Appl. Phys. Lett. **105**, 084105 (2014). https://doi.org/10.1063/1.4894092
35. A. Sgattoni, S. Sinigardi, L. Fedeli, F. Pegoraro, A. Macchi, Phys. Rev. E **91**, 013106 (2015). https://doi.org/10.1103/PhysRevE.91.013106
36. B. Eliasson, New J. Phys. **17**, 033026 (2015). https://doi.org/10.1088/1367-2630/17/3/033026
37. A. Macchi, F. Cornolti, F. Pegoraro, T.V. Liseikina, H. Ruhl, V.A. Vshivkov, Phys. Rev. Lett. **87**, 205004 (2001). https://doi.org/10.1103/PhysRevLett.87.205004
38. S. Atzeni, J. Meyer-ter-Vehn, *The Physics of Inertial Fusion* (Oxford University Press, 2004)
39. S. Chandrasekhar, *Hydrodynamic and hydromagnetic stability*, chap. X (Dover Publications, New York, 1981)
40. E. Ott, Phys. Rev. Lett. **29**, 1429 (1972). https://doi.org/10.1103/PhysRevLett.29.1429
41. F. Pegoraro, S.V. Bulanov, Phys. Rev. Lett. **99**, 065002 (2007). https://doi.org/10.1103/PhysRevLett.99.065002
42. V. Khudik, S.A. Yi, C. Siemon, G. Shvets, Phys. Plasmas **21**(1), 013110 (2014). https://doi.org/10.1063/1.4863845
43. M.G. Haines, Phys. Rev. Lett. **87**, 135005 (2001). https://doi.org/10.1103/PhysRevLett.87.135005
44. T.V. Liseykina, S.V. Popruzhenko, A. Macchi, New J. Phys. **18**, 072001 (2016). https://doi.org/10.1088/1367-2630/18/7/072001

Chapter 3
Laser Wakefield Accelerators: Plasma Wave Growth and Acceleration

Zulfikar Najmudin

Abstract Laser wakefield accelerators are now becoming established as indispensable laboratory tools. This is because of their ability to produce high energy electron beams in compact configurations. Here we outline the basic theory of laser wakefield acceleration. One dimensional plasma waves are considered and the differential equation for wakefield generation is derived. The case of wakefield generation at low laser intensity ($a_0 \ll 1$) is considered to demonstrate the relation between the laser driver and the generated plasma wave. Finally acceleration in the plasma wave is considered. The importance of non-linear and three dimensional effects on the plasma wave growth are discussed.

3.1 Laser Acceleration

Lasers that reach many terawatt, even petawatt, peak power are now common in research laboratories around the world. When focussed, the intensity of these laser systems can routinely exceed $1 \times 10^{18}\,\mathrm{Wcm}^{-2}$. At these intensities, the normalised (transverse) momentum of electrons oscillating in the laser field, $a \equiv p_\perp/mc$, can easily exceed 1. Hence, the electrons oscillate *relativistically* and one immediately envisions applications of lasers in particle acceleration.

However, since the motion is predominantly transverse to the laser propagation, the acceleration length is limited to the order of $a_0(\lambda/2\pi)$, where a_0 is the peak value of a for a laser of wavelength λ. This short acceleration length severely restricts the maximum energy gain. Ideally one would transform the transverse fields of the laser pulse into longitudinal fields that stay in phase with the laser pulse, and so travel at $\sim c$, making them ideal to accelerate relativistic particles. A way to do this is to use the laser pulse to drive a relativistic plasma wave, an idea first proposed by Tajima and Dawson [1].

Z. Najmudin (✉)
Blackett Laboratory, Department of Physics, The John Adams Institute for Accelerator Science, Imperial College London, London, UK
e-mail: z.najmudin@imperial.ac.uk

© Springer Nature Switzerland AG 2019   51
L. A. Gizzi et al. (eds.), *Laser-Driven Sources of High Energy Particles and Radiation*,
Springer Proceedings in Physics 231, https://doi.org/10.1007/978-3-030-25850-4_3

Consider a linear (one-dimensional) sinusoidal perturbation in the plasma electron density propagating along the z direction: $\delta n_e = \epsilon n_0 \sin(k_p z - \omega_p t)$, where ω_p is the classical plasma frequency, k_p is the waves wavenumber, and ϵ is the ratio of the amplitude of the wave to the initial plasma density n_0. For such a wave, moving with phase velocity close to the speed of light c, $k_p \approx \omega_p/c$. The associated electric field is given by Gauss' Law. For a wave with the maximum amplitude, $\epsilon = 1$, which is often called the "linear cold wavebreaking limit":

$$E = -\int -e(n_e - n_0)/\epsilon_0 \, dz = -(e n_0/\epsilon_0 k_p) \cos(k_p z - \omega_p t).$$

The maximum amplitude of the electric field is then,

$$E_0 = (e n_0/\epsilon_0 k_p) = mc\omega_p/e \approx 0.96\sqrt{n_e \, [\text{cm}^{-3}]} \, \text{Vcm}^{-1}. \tag{3.1}$$

Hence it should be possible to generate accelerating fields of order GV cm^{-1} for densities of around $1 \times 10^{18} \text{cm}^{-3}$. Compared with the fields produced by state-of-the-art conventional accelerators ($\approx \text{MV cm}^{-1}$). This gives a strong motivation for the development of plasma accelerators.

3.1.1 Plasma Wave Generation

The Ponderomotive Force: As noted above, the electric field of a laser is predominantly in the transverse direction, and this determines the primary motion of electrons in a laser field. However, as electrons oscillate they move away from the highest field regions of the laser to lower intensity regions where the restoring force is not as large. Therefore, over a cycle there is a resultant force in the direction away from regions of highest intensity. This forces is called the *ponderomotive force*, and can be considered as simply motion of any charged particle away from regions where the oscillation energy, which is called the ponderomotive potential, is highest.

We can derive an expression for the ponderomotive force by considering the equation of motion for electrons in the field:

$$\frac{d\mathbf{p}}{dt} = -e(\mathbf{E} + \mathbf{v} \times \mathbf{B})$$

The first term, which is a force that is linear to the strength of the laser field, will be considered explicitly. However the second term is inherently non-linear. In particular, as the electrons quiver in a laser field, their velocity is almost directly proportional to the transverse field, making the magnetic field term proportional to the field squared. We can simplify the effect of the non-linear term by writing $\mathbf{B} = \nabla \times \mathbf{A}$ and $\mathbf{v} = \mathbf{p}/\gamma m$:

$$\frac{d\mathbf{p}}{dt} = -e\,(\mathbf{v} \times \mathbf{B}) = -e\left(\frac{\mathbf{p}}{\gamma m} \times (\nabla \times \mathbf{A})\right)$$

To simplify matters, we take dimensionless units, such that $m \to 1, e \to 1, c \to 1$ and $p/mc \to p, eA/mc \to a$ etc. We can write the total rate of change in momentum in terms of temporal and convective parts on the left hand side, and also expand out the triple product on the right hand side to give:

$$\frac{\partial \mathbf{p}}{\partial t} + (\mathbf{v} \cdot \nabla)\mathbf{p} = -\frac{1}{\gamma}(\nabla_a(\mathbf{p} \cdot \mathbf{a}) - (\mathbf{p} \cdot \nabla)\mathbf{a}) \qquad (3.2)$$

The ∇_a on the first term on the right, signifies that the derivative should only be applied to the components of $\mathbf{a}$ after doing the dot product. The motion can be split into longitudinal and transverse components, and we can assume that the transverse motion is dominated by the laser oscillation, i.e.:

$$\mathbf{p} = \mathbf{p_x} + \mathbf{p_y} = \mathbf{p_x} + \mathbf{a}$$

Equation (3.2) can then be written:

$$\frac{\partial \mathbf{p}}{\partial t} = -\frac{1}{\gamma}\left(\frac{1}{2}\nabla a^2 - (\mathbf{p} \cdot \nabla)(\mathbf{a} - \mathbf{p})\right)$$

For the last term, we can write $\mathbf{a} - \mathbf{p} = \mathbf{p_x}$, then $(\mathbf{p} \cdot \nabla) \cdot \mathbf{p_x} = (\mathbf{a} + \mathbf{p_x} \cdot \nabla) \cdot \mathbf{p_x} \approx (\mathbf{p_x} \cdot \nabla) \cdot \mathbf{p_x}$. Here we've only considered the time averaged values of these terms. Since $\mathbf{a}$ oscillates more rapidly than the time scales we are interested in, the term proportional to $\mathbf{a}$ is assumed to time average to zero. By contrast the remaining two terms are always negative to (opposing) the gradient, so always point away from regions of high energy density ($\propto p^2$):

$$\frac{\partial \mathbf{p}}{\partial t} = -\frac{1}{2\gamma}\left(\nabla a^2 + \nabla p_x{}^2\right)$$

Since $\gamma = \sqrt{1 + p_x^2 + a^2}$, this can be written more concisely as,

$$\frac{\partial \mathbf{p}}{\partial t} = -mc^2\nabla\gamma \qquad \text{(Ponderomotive Force)}$$

where the constants were added to give the correct dimensions. $\partial \mathbf{p}/\partial t$ is the change of momentum at any given point in space, and hence is a useful for giving the force on fluid elements. The force term can be interpreted as being due to the gradient of a potential $U_p = \gamma mc^2$, which is the total energy of the electron due to either oscillations longitudinally or transversely due to the oscillating fields of a laser pulse. The latter allows a suitably shaped laser, in particular one with a longitudinal gradient, to be used to set a plasma in motion. Hence the ponderomotive force can be used in the generation of plasma waves in a wakefield accelerator, as we see in the next subsection.

Plasma wave generation: Electron plasma waves can be considered as electrostatic oscillations of a plasma which arise from the displacement of the electrons from quasi-neutrality with the background parent ions. For a simple treatment, it is sufficient to ignore magnetic fields in the plasma, and so we can describe the subsequent motion of the plasma using Gauss' Law, along with the fluid equations for the plasma electrons; continuity and motion:

$$\frac{\partial \mathbf{p}}{\partial t} = -e\mathbf{E} - mc^2\nabla\gamma \qquad \text{(motion)}$$

$$\nabla \cdot \mathbf{E} = -e(n_e - n_i)/\epsilon_0 \qquad \text{(Gauss)}$$

$$\frac{\partial n_e}{\partial t} + \nabla \cdot (n_e v) = 0 \qquad \text{(continuity)}$$

Of course we have implicitly included the effect of the laser's magnetic field in the non-linear force term, which as above are grouped together into the ponderomotive force term. As before, we can simplify by writing in normalised units, i.e. $e = 1$; $m_e = 1$; $c = 1$; $\epsilon_0 = 1$.

$$\frac{\partial \mathbf{p}}{\partial t} = -\mathbf{E} - \nabla\gamma \qquad \text{(motion)}$$

$$\nabla \cdot \mathbf{E} = -(n_e - n_i) \qquad \text{(Gauss)}$$

$$\frac{\partial n_e}{\partial t} + \nabla \cdot (n_e v) = 0 \qquad \text{(continuity)}$$

To simplify further we consider only longitudinal variations. As we will see later, the three dimensional nature of realistic plasma waves is important, but for now a one dimensional treatment is sufficient to understand many of the properties associated with plasma wave growth. So taking only variations in the x-direction, $E = E_x$; $p = p_x = \gamma\beta$. We also take $n_i = n_0$ since the ions are much heavier than the electrons and, on the timescales in which we are interested, can be considered to stationary.

$$\frac{\partial p}{\partial t} = -E - \frac{\partial\gamma}{\partial x} \qquad \text{(motion)}$$

$$\frac{\partial E}{\partial x} = (n_0 - n_e) \qquad \text{(Gauss)}$$

$$\frac{\partial n_e}{\partial t} + \frac{\partial}{\partial x}(n_e v) = 0 \qquad \text{(continuity)}$$

Since the plasma wave only evolves slowly in the frame in which the laser pulse is stationary, it is common to transform to that frame. Note this will be only a Gallilean transform for simplicity. Indeed, if we ignore the evolution of the laser pulse, then the plasma wave variables only depend on the spatial variable in this frame $\xi = x - t$:

$$\frac{\partial}{\partial t} = \frac{\partial \xi}{\partial t}\frac{\partial}{\partial \xi} = -\frac{\partial}{\partial \xi}; \qquad \frac{\partial}{\partial t} = \frac{\partial \xi}{\partial x}\frac{\partial}{\partial \xi} = \frac{\partial}{\partial \xi}$$

Hence we can convert the coupled differential equations above, which are functions of space and time, into equations only dependent on the spatial variable ξ in the frame in which things are 'quasistatic'. This assumption is thus called the Quasistatic Approximation:

$$-\frac{\partial p}{\partial \xi} = \frac{\partial \phi}{\partial \xi} - \frac{\partial \gamma}{\partial \xi} \tag{3.3}$$

$$\frac{\partial^2 \phi}{\partial \xi^2} = (n_e - n_0) \tag{3.4}$$

$$-\frac{\partial n_e}{\partial \xi} + \frac{\partial(n_e \beta)}{\partial \xi} = 0 \tag{3.5}$$

In the above equations, we have used that $E = -\frac{\partial \phi}{\partial \xi}$, and only consider the velocity in the moving frame. Note that we can directly integrate both (3.3) and (3.5). From (3.3):

$$\gamma - \beta\gamma - \phi = \text{constant} = 1$$
$$\rightarrow \quad \gamma(1 - \beta) = 1 + \phi \tag{3.6}$$

Here we assumed $\beta = \phi = 0$ initially to find the constant of motion. Similarly from (3.5):

$$n_e(1 - \beta) = \text{constant} = n_0$$
$$n_e = n_0/(1 - \beta) \tag{3.7}$$

In this case, initially $n_e = n_0$ and $\beta = 0$ as before. Interestingly, since $-1 < \beta < 1$, this implies $\frac{1}{2} < n_e < \infty$. Hence in 1D, the density in a plasma wave cannot fall below one half of its original density. We will find later that this is not the case in 3D. Note, that there is no such restriction on the maximum density in a plasma wave. But of course as the density peak grows in amplitude, so it must become narrower to conserve particle number. We can put (3.7) into Poisson's equation (3.4):

$$\frac{\partial^2 \phi}{\partial \xi^2} = n_0 \left(\frac{1}{1 - \beta} - 1 \right) \tag{3.8}$$

We can eliminate β from the above equation by noting, $\gamma^2 = 1 + p^2 + a^2 \rightarrow \gamma^2(1 - \beta^2) = 1 + a^2$. Using (3.6) above, we have $\gamma^2(1 - \beta)^2 = (1 + \phi)^2$. Dividing the two expressions gives:

$$\frac{1 + \beta}{1 - \beta} = \frac{1 + a^2}{(1 + \phi)^2}$$

A little bit of manipulation using this shows that we can rewrite (3.8) as,

$$\frac{\partial^2 \phi}{\partial \xi^2} = -\frac{1}{2}k_p{}^2 \left(\frac{1 + a^2}{(1 + \phi)^2} - 1 \right) \tag{3.9}$$

where now we included the constants $k_p{}^2 = \left(\dfrac{n_0 e^2}{\epsilon_0 m c^2} \right)$ to give an idea of the scale. As noted before, k_p is the wavenumber of the generated plasma wave and is only a function of the density n_0.

A more careful derivation that uses the group velocity of light frame, rather than the vacuum speed of light, gives the wakefield generation equation as [2, 3]:

$$\frac{1}{k_p{}^2} \frac{\partial^2 \phi}{\partial \xi^2} = \pm \frac{n_b}{n_0} + \gamma_p{}^2 \left\{ \beta_p \left[1 - \frac{(1 + a^2)}{\gamma_p{}^2(1 + \phi)^2} \right]^{-1/2} - 1 \right\} \tag{3.10}$$

Here the equation is written in terms of the quasistatic variable $\zeta = z - v_p t$ where the driver velocity is v_p. The equation has been generalised by including an extra space-charge term. The first term on the right describes the response to a particle beam driver of density n_b and is $+$ for an electron and $-$ for a positively charged driver. We note that this non-linear quasistatic 1D wakefield generation equation can be simplified to (3.9) by assuming the absence of beam drivers ($n_b = 0$) and taking the limit that the phase velocity of the wave $\gamma_p \gg 1$. Expanding the square brackets and also expanding $\beta_p = (1 - (1/\gamma_p{}^2))^{1/2}$ gives once again the non-linear laser wakefield driver equation (3.9) [3]:

$$\frac{1}{k_p{}^2} \frac{\partial^2 \phi}{\partial \xi^2} = \frac{1}{2} \left[\frac{(1 + a^2)}{(1 + \phi)^2} - 1 \right] \tag{3.11}$$

In the small amplitude limit $\phi \ll 1$, this can be further simplified to:

$$\left(\frac{\partial^2}{\partial \xi^2} + k_p{}^2 \right) \phi = \frac{1}{2} k_p{}^2 a^2 \tag{3.12}$$

Equations (3.10), (3.11), (3.12) can be solved directly to describe the wakefield generation in the quasistatic frame. Of the three, (3.12) is the most intuitive, as it shows that in the small amplitude limit, the wakefield generation is governed by a simple harmonic motion, that produces an oscillating solution driven by a term dependent on the ponderomotive force ($\propto a^2$). This can be most readily solved for arbitrary pulse shape by using Green's functions [4].

Regimes of wakefield generation: To elucidate the orgins of certain effects, rather than solving equations (3.10), (3.11), (3.12) directly, we start again from our base equations. In this instance, we start immediately in the quasistatic frame, and using normalised units:

$$\frac{\partial E}{\partial \zeta} = -n_1 \qquad \text{(Gauss' Law)}$$

$$\frac{\partial n_1}{\partial \zeta} = \frac{\partial (n_e \beta)}{\partial \zeta} \qquad \text{(Continuity)}$$

$$(1 - \beta)\frac{\partial \beta}{\partial \zeta} = eE + \frac{1}{\gamma}\frac{\partial (a^2)}{\partial \zeta} \qquad \text{(Motion)}$$

Here the normalisations of the variables are $\beta = v/c$ and $E = E/E_0$, where the normalisation of electric field $E_0 = mc\omega_p/e$ can be recognised as the linear cold wavebreaking limit given in the introduction. The density perturbation is defined by $n_e = n_0 + n_1$. Rather than solving for the potential, we instead solve for the electric field. So, rather than having one second order differential equation, we can solve two coupled first order equations. This can be solved using ode solvers such as those included in the Python Scipy package. Note that the ponderomotive force has now been split into terms dependent on the longitudinal and transverse momenta separately, or alternatively include the convective derivative explicitly in the equation of motion, which earlier was subsumed into the ponderomotive force.

To investigate the linear regime, we take $\beta \ll 1$, and $n_1 \ll n_0$, so that $n_e \approx n_0$. Our base equations become:

$$\frac{\partial E}{\partial \zeta} = -n_1 \qquad (3.13)$$

$$n_1 = n_0\beta \qquad (3.14)$$

$$\frac{\partial \beta}{\partial \zeta} = eE + \frac{\partial (a^2)}{\partial \zeta} \qquad (3.15)$$

The continuity equation (3.14) can be directly put into Gauss' Law (3.13) to leave coupled equations in E and β to solve. As an example, we can take a half-sinusoid laser intensity profile: $a^2 = a_0^2 \sin^2(\pi\zeta/L)$ for $-L < \zeta < 0$, and where $a = 0$ otherwise. Since the pulse is moving from left to right, for $\zeta > 0$, $n_1 = 0$. Figure 3.1 plots the corresponding electric field, fluid velocity in the quasistatic frame and electron density generated for a laser pulse with $a = 1.0$. The shape of the pulse (envelope of a) can be seen in red in the figure.

One can see that the laser pulses pushes electrons forward, which causes bunching ahead of the pulse and a corresponding positive electric field that eventually pulls electrons back. If the return is correctly timed, then the laser pulse gives a further push

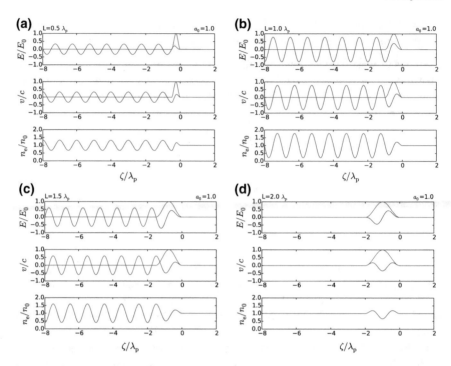

Fig. 3.1 Linear wakefield production: electric field E, fluid velocity v, and electron density n_e from one dimensional calculations in the quasistatic frame using the linear equations for $a_0 = 1$ and pulse lengths: **a** $L = \lambda_p/2$, **b** $L = \lambda_p$, **c** $L = 1.5\lambda_p$, **d** $L = 2\lambda_p$

to the electrons as they return, increasing the wave amplitude. Hence, the amplitude of the wakefield is optimised for $L = 1$ (note that all lengths are in units of λ_p) which is shown in Fig. 3.1b. This can be seen more clearly in Fig. 3.2, which plots the amplitude of the plasma wave as a function of laser pulse length. Here, the resonance at $L = \lambda_p$ is seen clearly. Note that we have defined L as the total length of the pulse, whilst experimentally it is more usual to talk about the full-width-half-maximum (fwhm) length of the pulse, in which case the resonance conditions is written:

$$L_{fwhm} = c\tau_{fwhm} = \lambda_p/2 \qquad (3.16)$$

where τ is the laser pulse length.

One can also see further resonances for pulse lengths at half-integers of the plasma wave wavelength, and also anti-resonances for integer values of L. These antiresonance can be seen more clearly in Fig. 3.1c, which plots the resulting wakefield for $L = 2\lambda_p$. The plasma wave behind the laser pulse is completely extinguished, but interestingly there is still a sizeable field within the duration of the laser pulse envelope. Figure 3.2 also plots the strength of the field within the first period as a function of laser pulse length, which except close to resonance, is larger than the size of the

Fig. 3.2 Linear wakefield
production: Plasma
wakefield amplitude as a
function of laser pulse length
L. In **blue**, the maximum
amplitude of the generated
longitudinal electric field,
including the field forced by
the laser envelope, **green** the
amplitude of the trailing
wakefield

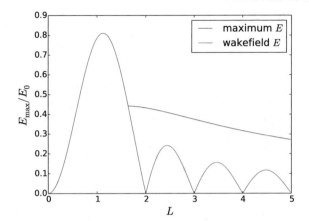

wakefield. It might be thought that it would be interesting to accelerate in a regime
where the pulse length is optimised to minimise the trailing wakefield, especially if
one is interested in only accelerating a single bunch. However in three dimensions,
it is not possible to completely extinguish the trailing field, due to the transverse
motion of plasma wave electrons. Hence it would only be possible to achieve this
kind of scenario by using a larger than necessary laser spot, which would at best be
inefficient.

In the linear regime, it is possible to analytically derive the plasma wave growth
in the quasistatic frame. For the case of a resonant pulse ($L = \lambda_p$):

$$E\left(\zeta\right) = \frac{1}{8}a_0{}^2 \left(k_p\zeta \cos\left(k_p\zeta\right) - \sin\left(k_p\zeta\right)\right) E_0 \qquad \text{for} - L < \zeta < 0.$$

The maximum field strength is then given, when $L = 1$ or $k_p\xi = 2\pi$, as $E_{\max} = (\pi/4)a_0{}^2$. Since the constants depend on the exact shape of the pulse and are of order
one, it is common to write $E_{\max} = a_0{}^2$ in the linear regime.

Clearly this solution has only limited validity. In particular, for $a_0 \gtrsim 1$, the density
variation can be $n_1 < -n_0$, leading to negative electron densities! To account for this,
we must treat the continuity equation properly,

$$n_e = (n_0 + n_1)\,\beta \tag{3.17}$$

As we noted before, $n_e = n_0 + n_1 = n_0/(1 - \beta)$. This prevents the density from
going below $n_e = \frac{1}{2}n_0$. As a result Gauss's Law becomes:

$$\frac{\partial E}{\partial \zeta} = -n_0 \frac{\beta}{1 - \beta} \tag{3.18}$$

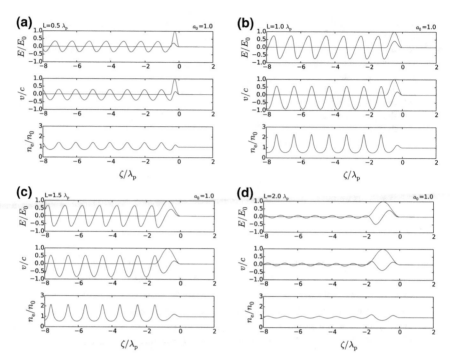

Fig. 3.3 Wakefield production with non-linear continuity: Electric field E, fluid velocity v, and electron density n_e for $a_0 = 1$ and pulse lengths: **a** $L = \lambda_p/2$, **b** $L = \lambda_p$, **c** $L = 1.5\lambda_p$. **d** $L = 2\lambda_p$

Solving this along with the equation of motion gives the wakefield solutions shown in Fig. 3.3. These show the characteristic flattening of the low density regions and peaking of the density spikes associated with non-linear plasma wave. We also note that as intensity a_0^2 is increased, the fields become less sinusoidal and more sawtooth in shape. This is characteristic of non-linear plasma waves. A final thing of note is that the plasma wave wavelength also becomes intensity dependent, such that for $L = 2\lambda_p$ the wake is not completely cancelled by the rear of the laser pulse. Since the minimum plasma density is restricted, it takes longer for the density minima to account for the now nonlinearly steepened density maxima.

Even for these weakly non-linear solutions, we note that $\beta \rightarrow 1$. To investigate intensities even higher than $a_0 \sim 1$, we need to consider a fully relativistic treatment of the plasma motion. To do this we can rewrite the equation of motion in terms of the fluid *momentum*, and restoring our convective term, which cannot now be considered to be small:

$$\frac{\partial E}{\partial \zeta} = -n_0 \frac{\beta}{1-\beta} \tag{3.19}$$

$$(1-\beta)\frac{\partial p}{\partial \zeta} = eE + \left(\frac{1}{\gamma}\right)\frac{\partial \left(a^2\right)}{\partial \zeta} \tag{3.20}$$

Here again $\beta = p/\gamma = p/\left(1 + p^2 + a^2\right)^{1/2}$. The effect of adding relativistic effects can be seen on the wakefield in Fig. 3.4, which plots the wakefield structure as a function of laser strength for a laser pulse of length $L = \lambda_p$. For $a \lesssim 0.4$, the wake structure resembles those given by the linear solutions. By $a_0 \approx 0.7$ discernible non-linearities can be observed, and for $a_0 = 1$, the wake field structure is significantly sawtoothed and lengthened. This is due to the greater inertia of the electrons performing the wake oscillation as they reach relativistic velocities. By $a_0 \approx 1.4$, the density peaks become almost δ-functions, leading to a near linear variation of the field with displacement. At higher field strengths, numerical simulations become unstable as the density spikes do indeed tend to infinity. In nature, this process becomes cataclysmic as well, resulting in the breakdown of the wave structure due to wavebreaking. Though wavebreaking limits the maximum strength of field that can be

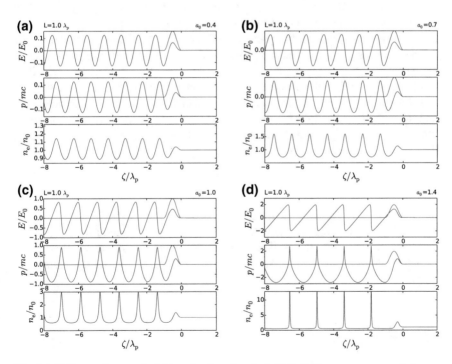

Fig. 3.4 Fully non-linear wakefield production: electric field E, fluid momentum p, and electron density n_e for pulse lengths $L = \lambda_p$ and **a** $a_0 = 0.4$, **b** $a_0 = 0.7$, **c** $a_0 = 1.0$, and **d** $a_0 = 1.4$

Fig. 3.5 Wakefield
amplitude: as a function of
laser strength using linear,
weakly nonlinear, and
relativistic treatments

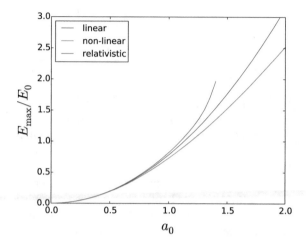

produced, this can be a simple way of injecting electrons into a wakefield, and is the
basis of most self-injection schemes [5–7].

Figure 3.5 plots the amplitude of the wakefield as a function of intensity including
both just the non-linearity due to the continuity equation, and also those including
relativistic effects. One sees that preventing full cavitation, as well as increasing the
plasma wave wavelength can lead to a reduction in the field strength at high a_0, but that
relativistic effects cause the E field to rise rapidly above linear calculations, whilst
also causing a rapid increase in plasma wave wavelength, and thus also the optimal
pulse length for resonant growth. Though it is difficult to calculate analytically the
field strength in the non-linear regime, it is possible to approximate the plasma wave
growth from a sharply rising pulse. In that case, the field strength is often quoted [3,
4] as:

$$E_{max}/E_0 \sim a_0^2/\left(1+a_0^2\right)^{1/2} \tag{3.21}$$

This scales as $\propto a_0$ for $a_0 > 1$ and, as found before, $\propto a_0^2$ in the linear regime.

And Fig. 3.6 plots the plasma wave wavelength as a function of intensity in the
linear regime along with the nonlinearity due to continuity and with continuity and
relativistic nonlinearities included. One can see that in the relativistic regime, the
plasma wave wavelength can scale as, $\lambda_{pw} \approx (1+a_0^2)^{1/2}\lambda_p$.

3.1.2 Energy Gain

Linear Regime and dephasing We now consider the energy gain of electrons in the
wakefield. Again we start with linear waves to give some physical insight into the
important characteristics of the acceleration process, before considering corrections

Fig. 3.6 Wakefield length: as a function of laser strength using linear, weakly nonlinear, and relativistic treatments

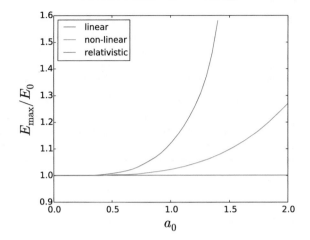

required for acceleration in non-linear wakes. For linear wake of maximum amplitude $n_1 = n_0$, in the quasistatic frame the density perturbation becomes;

$$n_1 = n_0 \cos k_p \xi$$

We can integrate twice, firstly using Gauss' Law to find the electric field, and then again to find the corresponding potential:

$$E = -\frac{e n_0}{\epsilon_0 k_p} \sin k_p \xi$$

$$\phi = -\frac{e n_0}{\epsilon_0 k_p^2} \cos k_p \xi$$

The maximum energy gain for an electron in this potential is then $W = -2e(\phi_{\min} - \phi_{\max}) = 2\frac{n_0 e^2}{\epsilon_0 k_p^2} = 2\frac{n_0 e^2}{\epsilon_0 m}\frac{mc^2}{\omega_p^2} = 2mc^2 \approx 1$ MeV. Clearly this is not too impressive, but we have not considered that the wave is moving. First, we find the acceleration in the relativistically boosted frame of the plasma wave. The Lorentz transformations tell us that the longitudinal field is unaffected: $E' = E$, but the length of the wave is elongated by the relativistic transformation [1]. Hence the wavenumber of the wave is reduced as: $k_p \rightarrow k_p' = k_p/\gamma_{ph}$, where γ_{ph} is the phase velocity associated with the plasma wave. So in the boosted frame, the electric field and potential transform as:

$$E' = -\frac{e n_0}{\epsilon_0 k_p} \sin\left(k_p' \xi'/\gamma_{ph}\right)$$

$$\phi' = -\gamma_{ph}\frac{e n_0}{\epsilon_0 k_p^2} \cos\left(k_p' \xi'/\gamma_{ph}\right)$$

Hence, the energy gain in the boosted frame is $W = 2\gamma_{ph}mc^2$. This is fine in 1D, but in three dimensions the plasma wave only has a focussing force over half of its length. More prosaically, the potential of a linear wave has a hill and valley structure, (strictly we are talking about the negative of the potential here since electrons have a negative charge). If an electron begins at the top of a valley, then it will always fall to the bottom. However if it starts at the top of the hill, there is no guarantee that it will fall in the direction in which the plasma wave is moving—along the axis. Hence acceleration is only guaranteed over one half of the plasma wave's total length, and the maximum energy gain is also reduced by half. So the energy gain and the corresponding momentum (assuming $\gamma_{ph} \gg 1$) in the boosted frame are:

$$W' = \gamma_{ph}mc^2$$
$$p' \simeq \gamma_{ph}mc$$

We can use the Lorentz transformations to find the energy gain in the laboratory.

$$\begin{pmatrix} iW \\ cp \end{pmatrix} = \begin{pmatrix} \gamma_{ph} & i\beta_{ph}\gamma_{ph} \\ -i\beta_{ph}\gamma_{ph} & \gamma_{ph} \end{pmatrix} \begin{pmatrix} iW' \\ cp' \end{pmatrix}$$
$$iW = \gamma_{ph}W' + i\beta_{ph}\gamma_{ph}cp'$$

For $\gamma_{ph} \gg 1$, $\beta_{ph} \to 1$,

$$W \simeq \gamma_{ph}^2mc^2 + \gamma_{ph}^2mc^2 = 2\gamma_{ph}^2mc^2 \qquad (3.22)$$

Hence we can see that the factor of two comes from the contribution from the momentum part of the Lorentz transformation and not from being able to accelerate over the whole length of the plasma wave.

The phase velocity of the plasma wave v_{ph} is determined by the speed at which the laser travels through the plasma, i.e. the laser group velocity, v_g. (NB a cold plasma wave has no group velocity in itself, the plasma wave structure will continue to oscillate in place without moving!) Hence, starting from the dispersion relation for electromagnetic waves in plasma: $c^2k^2 = \omega^2 - \omega_p^2$, then $\omega = ck/\eta$, where the plasma refractive index $\eta = \sqrt{1 - \omega_p^2/\omega^2}$. Then,

$$v_{ph} = v_g = \frac{\partial\omega}{\partial k} = c\sqrt{1 - \omega_p^2/\omega^2} = \eta c \quad \to \quad \beta_{ph} = \eta$$
$$\gamma_{ph} = \left(1 - \beta_{ph}^2\right)^{-1/2} = \left(1 - \left(1 - \frac{\omega_p^2}{\omega^2}\right)\right)^{-1/2} = \left(\frac{\omega_0}{\omega_p}\right) = \left(\frac{n_{cr}}{n_0}\right)^{1/2} \qquad (3.23)$$

So finally, we can give more generally the energy gain from the plasma accelerator in terms of the ratio of initial density to critical density n_0/n_{cr} and the ratio of the plasma wave amplitude to the ambient density, $\epsilon = \delta n/n_0$.

$$W = \epsilon \cdot 2mc^2 \cdot \frac{n_{cr}}{n_0} \qquad (3.24)$$

ϵ is thus also the ratio of the maximum field strength to the linear cold wavebreaking limit which in the linear regime was found above to be $\epsilon = E/E_0 \simeq a_0^2$, which gives in the linear regime, $W = 2mc^2(n_{cr}/n_0)a_0^2$. As one can see, surprisingly the maximum energy gain actually increases with reduced plasma density, even though the maximum field strength decreases $\propto \sqrt{n_0}$ as noted in (3.1). This is purely due to the increased phase velocity of the plasma wave, so that it takes the electron, which is now travelling very close to c, longer before it outruns the accelerating part of the plasma wave. This process is called dephasing, and the length over which the electron dephases is L_{deph}. We found earlier that the acceleration length in the boosted frame is half the plasma wave length $L' = \pi/k_p' = \gamma_{ph}\pi/k_p$. The time taken to traverse the wave is then $t' \simeq L'/c$. Lorentz transforming,

$$\begin{pmatrix} ict \\ L \end{pmatrix} = \begin{pmatrix} \gamma_{ph} & i\beta_{ph}\gamma_{ph} \\ -i\beta_{ph}\gamma_{ph} & \gamma_{ph} \end{pmatrix} \begin{pmatrix} ict' \\ L' \end{pmatrix}$$

$$L_{\mathrm{deph}} = \gamma_{ph}cL'/c + \gamma_{ph}L' = 2\gamma_{ph}^2\pi/k_p = \gamma_{ph}^2\lambda_p$$

where we have written $\lambda_p = 2\pi/k_p = 2\pi c/\omega_p$, which is the wavelength of a relativistic plasma wave. We could have found this directly by noting that this is the acceleration over half of the length of plasma wave length multiplied by the Doppler factor of $2\gamma_{ph}^2$. We can write the dephasing length in terms of density using the expression for γ_{ph} in (3.23),

$$L_{\mathrm{deph}} = \frac{n_{cr}}{n_0}\frac{\omega_0}{\omega_p}\frac{2\pi c}{\omega_0} = \left(\frac{n_{cr}}{n_0}\right)^{3/2}\lambda_0 \qquad (3.25)$$

where λ_0 is the laser wavelength. We note that whilst the energy gain increases linearly with the inverse of the density, the dephasing length increases more quickly ($\propto n_0^{-3/2}$). Hence, the average acceleration strength decreases, and one trades high energy gain with less compact size. Nevertheless, even for relatively low densities of around $n_0 = 1 \times 10^{14}\mathrm{cm}^{-3}$, $E_{\mathrm{acc}} \simeq 1\,\mathrm{GVm}^{-1}$, which is still an order of magnitude larger than established acceleration techniques. Below these densities, there is little point to use plasma based accelerators.

Laser depletion: A final important consideration is that the driver energy should be sufficient to grow a plasma wave over the dephasing length to be able to obtain the maximum predicted energy at a given density. We can find the length over which the laser is depleted, by equating the energy given to the plasma wave to the energy in the laser pulse. In 1D, the (areal) energy density in the plasma wave is $\frac{1}{2}\epsilon_0 E^2$. Taking $E = \epsilon E_0 \sin(k_p\xi)$ as before, then per plasma wave wavelength,

$$U_{\mathrm{pw}} = \frac{1}{2}\int_0^{\lambda_p} \epsilon_0 E^2 \mathrm{d}\xi = \frac{1}{2}\epsilon_0 \int_0^{2\pi/k_p} \epsilon^2 E_0^2 \sin^2(k_p\xi)\mathrm{d}\xi = \frac{1}{4}\epsilon_0\epsilon^2 E_0^2\lambda_p$$

Substituting for the linear cold-wavebreaking limit E_0 gives an energy loss of $U_{pw} = \frac{1}{4}\epsilon^2 n\,mc^2\lambda_p$ per plasma wave wavelength and,

$$U_{pw} = \frac{1}{4}\epsilon^2 n\,mc^2 \qquad (3.26)$$

per unit length. Similarly the (areal) energy density in the laser pulse is on average $\frac{1}{2}\epsilon_0 E_l^2$, where we have noted that the electric and magnetic parts of the field equally contribute $\frac{1}{2}\epsilon_0 E_l^2$, but there is another factor of $\frac{1}{2}$ from the time-averaging of the oscillating field. For a laser pulse with an envelope $a = a_0 \sin(\pi\xi/c\tau)$ for $0 < \xi < c\tau$, the energy in the laser field is then:

$$U_1 = \frac{1}{2}\int_0^{c\tau} \epsilon_0 E_l^2 d\xi = \frac{1}{2}\epsilon_0 \int_0^{c\tau} \left(\frac{m\omega_0 c}{e}\right)^2 a_0^2 \sin^2(\pi\xi/c\tau) d\xi = \frac{1}{4}n_{cr}mc^2 a_0^2 c\tau \qquad (3.27)$$

Dividing by (3.26) gives the length over which this laser energy is depleted as:

$$L_{depl} = (n_{cr}/n_0)\, a_0^2\, c\tau/\epsilon^2 = (n_{cr}/n_0)\, c\tau/a_0^2, \qquad (3.28)$$

where we have approximated $\epsilon \simeq a_0^2$ for the linear regime. As noted above, to grow the plasma wave optimally in the linear regime, we choose $c\tau = \lambda_p$ and so in this case:

$$\text{For} \quad c\tau = \lambda_p : \qquad L_{depl} = (n_{cr}/n_0)\, \lambda_p/a_0^2 = L_{deph}/a_0^2 \qquad (3.29)$$

So fortunately, in the linear regime depletion is never shorter than the dephasing length, and only becomes an issue as $a_0 \to 1$. Indeed, this can be a useful feature as it means that acceleration is terminated once dephasing is reached. It may be thought that the depletion of the laser pulse with increased length might be an impediment to quasistatic acceleration with a laser pulse. Clearly it would not be ideal for the field strength to decrease as the laser and electron beam propagate. In reality, pulse compression means that the laser power can remain relatively unchanged, meaning the plasma wave amplitude can be quite consistent over the whole acceleration length [8]. Indeed, at even higher initial laser strength, this pulse compression can be become severe enough to cause amplification of the laser power [9]. Hence for controllable acceleration it is thought best to remain in the quasi non-linear regime $1 \lesssim a_0 \lesssim 2$.

Non-linear and 3D considerations The simple scalings that we have presented until now prove to be very useful in determining the requirements of experiments. However, in reality it is common to run experiments at high ($a_0 \sim 1$) or very high field strengths ($a_0 > 1$), not only to obtain very high field strengths, and thus compact acceleration, but also to be able to inject electrons directly from the plasma [5–7].

Luckily rather than rederiving all of the expressions from above, it is possible to recast the equations we derived above from the linear regime, but taking account of the intensity dependencies of the main parameters that affect the acceleration. In particular, the energy gain depends on the plasma wave amplitude, length and phase velocity all become dependent on the intensity. We can recall the dependence on intensity for each. For the amplitude, we found: $E/E_0 \approx a_0^2/\sqrt{1 + a_0^2}$. Therefore for high fields, $E \approx a_0 E_0$. Similarly we noted that the plasma wave wavelength increases by $\lambda_p^* = \gamma_\perp \lambda_p \approx a_0 \lambda_p$. One may think that with both of these effects combined, that the energy gain would continue to scale as a_0^2. However, as we also noted, this 1D non-linear regime is extremely difficult to attain since it needs very large laser focal spots of high intensity, thus implying the need for extreme laser energies.

Also, the waves phase velocity in the non-linear regime is reduced from the linear group velocity given above. This is because the front of the laser pulse is found to be etched away due to the strong local laser depletion there. From simulation, it is has been found that the velocity at which this etching of the laser pulse occurs in the pulse frame is $v_{etch} = c(\omega_p^2/\omega_0^2)$. The depletion length becomes:

$$L_{depl} \approx \left(\omega_0^2/\omega_p^2\right) c\tau$$

which again for an optimal duration laser pulse, $c\tau \approx \lambda_p$ gives $L_{depl} \approx L_{deph}$. The etching reduces the effective group velocity of the pulse and the resultant plasma wave phase velocity:

$$v_{ph}/c = \left(v_g - v_{etch}\right)/c \approx 1 - \frac{3}{2}\left(\omega_p^2/\omega_0^2\right) \tag{3.30}$$

And thus $\gamma_{ph} = \frac{1}{\sqrt{3}}\left(\omega_0/\omega_p\right)$. This reduces the dephasing length to:

$$L_{deph} = \frac{2}{3}\left(\omega_0/\omega_p\right)^2 \lambda_p = \frac{2}{3}\left(n_{cr}/n_0\right)^{3/2} \lambda_0 \tag{3.31}$$

One can see that in the weakly non-linear regime, the pulse front etching opposes the non-linear steepening of the plasma wave, and as a result that energy gain is not much greater than given by the linear scalings for modest values of a_0. We also note that the laser pulse becomes fully 'etched away' after a distance:

$$L_{depl} = \frac{c\tau}{v_{etch}/c} = c\tau\left(\frac{n_{cr}}{n_0}\right) = \frac{c\tau}{\lambda_p}\left(\frac{n_{cr}}{n_0}\right)^{3/2} \lambda_0$$

One sees that depletion is thus similar to that for the non-linear expression for $a_0 = 1$.

Of course we noted that in 1D full cavitation of the plasma waves is not possible. However in 3D, it becomes optimal to focus the laser pulse to something of the order of the plasma wave wavelength in all three dimensions. Now the plasma wave looks like a bubble, with a high density wall surrounding a central void, and indeed this regime is often colloquially termed the 'bubble' regime. We can make simple estimates for the acceleration in this regime by noting that the bubble size grows until the space charge field is approximately the size of the ponderomotive force. Since the field of a spherically symmetric collection of charge of density n is $E = \frac{1}{3}(en_0/\epsilon_0)r$, then:

$$-\frac{1}{3}n\frac{e^2}{e_0}r = -mc^2\nabla\gamma \approx \frac{-mc^2\sqrt{1+a_0^2}}{r}$$

which gives,

$$r = \sqrt{3a_0}\left(\frac{c}{\omega_p}\right) \qquad \text{for} \qquad a_0 \gg 1$$

which implies that the maximum field strength is

$$E_{\max} = \sqrt{\frac{a_0}{3}}\left(\frac{mc\omega_p}{e}\right) \qquad \text{for} \qquad a_0 \gg 1$$

One notes that as the charge seperation occurs in multiple directions, the scaling of maximum electric field with increasing a_0 is slower than in the 1D non-linear regime. Since the average field is only half of this maximum value, and the acceleration is only over half the size of the bubble, so that the acceleration length equals r, then the energy gain in the 3D non-linear regime is:

$$W_{\max} = 2\gamma_{ph}^2 \cdot \frac{1}{2}E_0 \cdot r = \frac{2}{3}\left(\omega_0/\omega_p\right)^2 a_0 mc^2$$

One can see that one has to operate at laser strengths significantly greater than $a_0 = 1$ to see enhanced acceleration in this regime, due to the reduction in plasma wave velocity associated with the pulse front etching.

Formula summary: We summarise the results presented in this manuscript in the table below, as adapted from the original scalings presented by Lu et al. [10]. Table 3.1 presents the linear scalings as function of (n_{cr}/n_0) for a wave of amplitude $\epsilon = 1$ to emphasise that these parameters determine the properties of the laser wakefield accelerator in all the regimes. Table 3.2 then presents the scaling of the laser wakefield parameters in terms of those defined in Table 3.1 for the different regimes, now emphasising the dependence on laser strength a_0 too.

We use the scalings derived above to give an indication of the laser parameters required for a laser wakefield accelerator. For example to obtain a GeV energy gain in the linear regime with $a_0 = 1$, we find that using $W_{\max} = 2mc^2(n_{cr}/n_0) \approx (n_{cr}/n_0)\,\text{MeV} = 1\,\text{GeV}$. Therefore, $n_0 \approx 10^{-3}n_{cr}$. Much of the advance in laser wakefield acceleration has been made due to the advances in Ti:sapphire lasers, which support ultrashort laser pulses, and operate around $\lambda = 800\,\text{nm}$, for which

Table 3.1 Linear scalings for laser wakefield accelerator of amplitude $\epsilon = 1$, emphasising scaling with initial plasma density n_0

Parameter	w_0	λ_{pw}	γ_{ph}	L_{deph}	E_{max}	W_{max}
Symbol	w_0	λ_p	γ_{p0}	L_0	E_0	W_0
Scaling	$\sqrt{\dfrac{n_0}{n_{cr}}}\lambda_0$	$\sqrt{\dfrac{n_0}{n_{cr}}}\lambda_0$	$\sqrt{\dfrac{n_{cr}}{n_0}}$	$\left(\dfrac{n_{cr}}{n_0}\right)^{3/2}\lambda_0$	$\sqrt{\dfrac{n_0}{n_{cr}}}\left(\dfrac{mc\omega_0}{e}\right)$	$2\left(\dfrac{n_{cr}}{n_0}\right)mc^2$

Table 3.2 Relative scalings for laser wakefield parameters in various regimes, along with limits of validity of each approximation

Regime	a_0	w_0/λ_p	λ_{pw}/λ_p	γ_{ph}/γ_{p0}	L_{deph}/L_0	L_{depl}/L_0	E_{max}/E_0	W_{max}/W_0
Linear:	< 1	1	1	1	1	$\left(\dfrac{1}{a_0^2}\right)\left(\dfrac{c\tau}{\lambda_p}\right)$	a_0^2	a_0^2
1D NL	> 1	~ 1	$\sqrt{1+a_0^2}$	$\dfrac{1}{\sqrt{3}}$	$\dfrac{\sqrt{1+a_0^2}}{3}$	$\left(\dfrac{c\tau}{\lambda_p}\right)$	$\dfrac{a_0}{\sqrt{1+a_0^2}}$	$\dfrac{2}{3}\dfrac{a_0^4}{\sqrt{1+a_0^2}}$
3D − NL	> 2	$\dfrac{1}{2\pi}\sqrt{3a_0}$	$\dfrac{1}{2\pi}\sqrt{3a_0}$	$\dfrac{1}{\sqrt{3}}$	$\dfrac{\sqrt{a_0}}{3\sqrt{3}}$	$\left(\dfrac{c\tau}{\lambda_p}\right)$	$\sqrt{\dfrac{a_0}{3}}$	$\dfrac{1}{3}a_0$

$n_{cr} \approx 1.7 \times 10^{21}\,\text{cm}^{-3}$. This implies $n_0 \approx 1.7 \times 10^{18}\,\text{cm}^{-3}$. For a plasma wave amplitude $\epsilon = 1$, $E_{max} = E_0 \approx 125\,\text{GVm}^{-1}$. The laser should be focussed to spot size $w_0 \approx \lambda_p = (n_{cr}/n_0)^{1/2}\lambda_0 = 25\,\text{mm}$, and the corresponding pulse length should be $c\tau_{fwhm} \approx \lambda_p/2$, giving $\tau_{fwhm} \approx 40\,\text{fs}$. Since it is important to be able to attain such short pulse lengths to efficiently grow the plasma wave, it can be seen why it was not until the development of Ti:sapphire lasers, which allowed such short pulses to be amplified at high power, that laser wakefield accelerators could be operated efficiently [5–7]. The corresponding dephasing (and depletion) length is $L_0 \approx 2.5\,\text{mm}$, demonstrating the compactness of the accelerator.

Finally we note that the energy in a laser pulse which is $\approx \lambda_p$ in spot size and length is given by:

$$E_l = \epsilon_0 E^2 \cdot \pi w_0^2 \cdot c\tau = n_{cr} \cdot mc^2 \cdot a_0^2 \cdot \lambda_p^3 = n_{cr}mc^2\lambda_0^3 \cdot a_0^2 \cdot (n_{cr}/n_0)^3$$

Calculating the constants, we find that $E_l \approx 70 \cdot a_0^2 \cdot (n_{cr}/n_0)^{3/2}\text{mJ}$. Again we note the scaling on laser energy requirement is faster than the gain in energy per stage, which motivates the use of multiple stages to reach high energy efficiently [11, 12]. Using this scaling for a GeV acceleration stage, we find a minimum energy of around 2.5 J, which is close to the value reported for the most efficient acceleration to this energy level [13].

References

1. T. Tajima, J.M. Dawson, Laser electron accelerator. Phys. Rev. Lett. **43**, 267–270 (1979)
2. V.I. Berezhiani, I.G. Murusidze, Interaction of highly relativistic short laser pulses with plasmas and nonlinear wake-field generation. Phys. Scr. **45**, 87 (1992)
3. S.V. Bulanov, V.I. Kirsanov, A.S. Sakharov, Excitation of ultrarelativistic plasma waves by pulse of electromagnetic radiation. JETP Lett. **50**, 198 (1989)
4. E. Esarey, C. Schroeder, W. Leemans, Physics of laser-driven plasma-based electron accelerators. Rev. Mod. Phys. **81**, 1229 (2009)
5. S.P.D. Mangles, C.D. Murphy, Z. Najmudin, A.G.R. Thomas, J.L. Collier, A.E. Dangor, E.J. Divall, P.S. Foster, J.G. Gallacher, C.J. Hooker, D.A. Jaroszynski, A.J. Langley, W.B. Mori, P.A. Norreys, F.S. Tsung, R. Viskup, B.R. Walton, K. Krushelnick, Monoenergetic beams of relativistic electrons from intense laser-plasma interactions. Nature **431** (2004)
6. C.G.R. Geddes, C. Toth, J. Van Tilborg, E. Esarey, C.B. Schroeder, D. Bruhwiler, C. Nieter, J. Cary, W.P. Leemans, High-quality electron beams from a laser wakefield accelerator using plasma-channel guiding. Nature **431**, 538 (2004)
7. J. Faure, C. Rechatin, A. Norlin, A. Lifschitz, Y. Glinec, V. Malka, Controlled injection and acceleration of electrons in plasma wakefields by colliding laser pulses. Nature **444**, 737 (2006)
8. J. Schreiber, C. Bellei, S.P.D. Mangles, C. Kamperidis, S. Kneip, S.R. Nagel, C.A.J. Palmer, P.P. Rajeev, M.J.V. Streeter, Z. Najmudin, Complete temporal characterization of asymmetric pulse compression in a laser wakefield. Phys. Rev. Lett. **105**, 235003 (2010)
9. M.J.V. Streeter, S. Kneip, M.S. Bloom, R.A. Bendoyro, O. Chekhlov, A.E. Dangor, A. Dpp, C.J. Hooker, J. Holloway, J. Jiang, N.C. Lopes, H. Nakamura, P.A. Norreys, C.A.J. Palmer, P.P. Rajeev, J. Schreiber, D.R. Symes, M. Wing, S.P.D. Mangles, Z. Najmudin, Observation of laser power amplification in a self-injecting laser wakefield accelerator. Phys. Rev. Lett. **120**, 254801 (2018)
10. W. Lu, M. Tzoufras, C. Joshi, F. Tsung, W. Mori, J. Vieira, R. Fonseca, L. Silva, Generating multi-GeV electron bunches using single stage laser wakefield acceleration in a 3D nonlinear regime. Phys. Rev. Spec. Top. Accel. Beams **10**, 061301 (2007)
11. W. Leemans, E. Esarey, Laser-driven plasma-wave electron accelerators. Phys. Today **62**, 44 (2009)
12. S. Steinke, J. Van Tilborg, C. Benedetti, C.G.R. Geddes, C.B. Schroeder, J. Daniels, K.K. Swanson, A.J. Gonsalves, K. Nakamura, N.H. Matlis, B.H. Shaw, E. Esarey, W.P. Leemans, Multistage coupling of independent laser-plasma accelerators. Nature **530**, 190 (2016)
13. W.P. Leemans, B. Nagler, A.J. Gonsalves, C. Tth, K. Nakamura, C.G.R. Geddes, E. Esarey, C.B. Schroeder, S.M. Hooker, C. Toth, GeV electron beams from a centimetre-scale accelerator. Nat. Phys. **2**, 696 (2006)

Chapter 4
LWFA Electrons: Staged Acceleration

Masaki Kando

Abstract This lecture focuses on the topic of staged laser wakefield acceleration together with the fundamentals of beam physics, which is necessary for discussing the topic. Staged acceleration is considered to accelerate an electron energy or to control an electron beam in phase space.

4.1 Introduction

Laser wakefield acceleration (LWFA) is conceived as promising for compact electron accelerators after the breakthrough achievements done over 20 years. As in other technologies in the early stage of the development LWFA had very poor energy gain and a broad energy distribution, however it has been proved to generate quasi-monoenergetic(<1%), high-energy (>1 GeV), ultrashort (<2 fs rms), low emittance (<1 mm-mrad), low-divergence (~1 mrad) beams although all these parameters have not yet confirmed simultaneously. These results were obtained in a rather simple setup; a single intense laser pulse was focused onto a gas target. Gas targets used were supersonic gas-jets, plasma waveguides, structured density targets, and so on. To develop LWFA further, it is natural to introduce the staging concept, which was used in radio-frequency accelerators.

Staging of acceleration structures is actually not well defined in LWFA. In a narrow sense, a stage is a structure driven by a dedicated laser pulse for a special intention such as acceleration, bunching or debunching. In a broad sense, a stage is a unit structure that can be either using the same laser as other stages or a dedicated laser pulse for the stage. This difference can be seen in the review section of staged laser wakefield acceleration. In this lecture I adopt a broader definition of the terminology.

M. Kando (✉)
Kansai Photon Science Institute, National Institutes for Quantum and Radiological Science and Technology, 8-1-7 Umemidai, Kizugawa, Kyoto 619-0215, Japan
e-mail: kando.masaki@qst.go.jp
URL: http://www.kansai.qst.go.jp/kpsi-en/

© Springer Nature Switzerland AG 2019
L. A. Gizzi et al. (eds.), *Laser-Driven Sources of High Energy Particles and Radiation*,
Springer Proceedings in Physics 231, https://doi.org/10.1007/978-3-030-25850-4_4

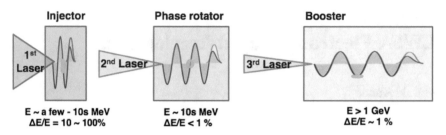

Fig. 4.1 A three-stage concept adopted in the ImPACT program in Japan. Three different laser pulses (1 J/20 fs, 2 J/50 fs, and 10 J/100 fs) are used for an injector, phase rotator, and booster stages

There are several reasons to adopt staging in accelerators. One is to accelerate electrons further because one stage is limited by depletion of the driving RF source or laser pulse. The other rather positive reason is to manipulate electrons properly to obtain higher quality electrons or conditioning of the electron beam for further acceleration. A good example is a linear accelerator (linac) for X-ray free-electron lasers. A DC electron gun produces an electron bunch with very low energy spread but a long bunch length. This electron is captured and bunched in a sub-harmonic buncher (SHB), which is operated at a lower frequency (sub-harmonic) to compress electron bunch length for the main linac. In SACLA the several RF frequencies are employed to obtain tens of femtosecond electron bunches [1].

Another example is seen in a staged LWFA project in Japan as depicted in Fig. 4.1 [2]. In the project three laser pulses are being considered to operate three stages efficiently. The first laser pulse, which is short (20 fs) and has a pulse energy of 1 J, is focused tightly in a high density gas jet. This injector stage produces low energy, high charge electrons via wave breaking. A selected portion of the electrons are further injected to the second stage, where the beam is compressed in time. In the second stage a longer pulse duration of 50 fs is utilized for a larger acceptance. The third stage is used to accelerate the injected electron up to 1 GeV with a 10-J, 100-fs laser pulse with a linear or quasi-linear laser intensity with a few centimeters long waveguide.

An European project on XFEL EuPRAXIA is also in progress, which widely investigates possible solutions to realize compact XFELs including laser based, combination of RF linacs and laser accelerators, etc [3].

Other projects are also working to construct LWFA based FELs in China and the U.S.A., etc.

4.2 Basic Concepts of Beam Physics

Here we introduce some basic useful concepts from beam physics or accelerator physics, which will be needed to discuss the following sections in the paper [4–7].

4.2.1   Transverse Beam Dynamics

Let us consider the situation that an electron beam is moving in a static magnetic field in a Cartesian coordinate system depicted in Fig. 4.2. The charged particle is deflected in x direction with a nominal curvature radius of ρ_0 in the magnetic field of B_z. We consider a displacement x in the transverse direction from the central orbit (denoted 0) here. The equation of motion of a charged particle with a mass of m and a charge of qe is

$$m\gamma \frac{d^2 x}{dt^2} = \frac{m\gamma v^2}{\rho_0 + x} + qev B_z(x), \tag{4.1}$$

where γ is the Lorenz factor of the particle, v is the velocity.
    The equation can be simplified by using the relations

$$\frac{dx}{dt} = \frac{dx}{ds}\frac{ds}{dt} = v\frac{dx}{ds} \tag{4.2}$$

to

$$\frac{d^2 x}{dt^2} = v^2 \frac{dx}{ds}. \tag{4.3}$$

If the displacement x is small and $B_z(x)$ is small, then the following two approximations can be made:

$$\frac{1}{\rho_0 + x} \approx \frac{1}{\rho_0}\left(1 - \frac{x}{\rho_0}\right) \tag{4.4}$$

$$B_z(x) \approx B_z(0) + \frac{dB_z}{dx}x. \tag{4.5}$$

Also in the central orbit the centrifugal force and Lorentz force is balanced:

$$\frac{m\gamma v^2}{\rho_0} + qev B_z(0) = 0. \tag{4.6}$$

Fig. 4.2 Charged particle motion in the horizontal plane around a central orbit in a magnetic field

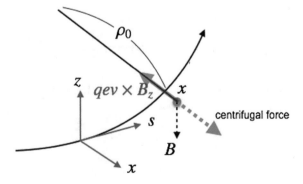

Fig. 4.3 Charged particle motion in a vertical plane around a central orbit in a magnetic field

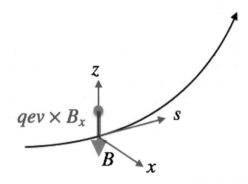

Thus (4.1) simplifies to

$$m\gamma v^2 \frac{d^2x}{ds^2} = -\frac{m\gamma v^2}{\rho_0^2}x + qev\frac{dB_z}{dx}x. \tag{4.7}$$

Here we introduce a field gradient index as

$$n := -\frac{\rho_0}{B_z(0)}\frac{dB_z}{dx} \tag{4.8}$$

then (4.7) is expressed as

$$\frac{d^2x}{ds^2} = -\frac{1-n}{\rho_0^2}x. \tag{4.9}$$

This is an equation of motion in the horizontal (x) direction.

Next let us consider the vertical motion of a charged particle as depicted in Fig. 4.3.

An equation of motion in the vertical direction (perpendicular to the bending plance) is

$$m\gamma\frac{d^2z}{dt^2} = -qevB_x(z). \tag{4.10}$$

Similar to the horizontal plane we convert the variable t to s using (4.3) and a linearization of $B_x(z)$

$$B_x(z) \approx B_x(0) + \left.\frac{dB_x(z)}{dz}\right|_{z=0}z, \tag{4.11}$$

where $B_x(0) = 0$ in the median plane. Using $\nabla \times \mathbf{B} = 0$ we obtain

$$\frac{dB_x}{dz} - \frac{dB_z}{dx} = 0. \tag{4.12}$$

Using (4.8), (4.11) and (4.12) the equation of motion in the vertical direction (4.10) simplifies to

$$\frac{d^2z}{ds^2} = -\frac{n}{\rho_0^2}z. \tag{4.13}$$

One can understand the stability of the charged particle in a magnetic field by using the two equations of motion (4.9) and (4.13). If $n < 1$ is satisfied, the charged particle can be confined in some region. This condition is called the *weak focusing condition*.

$$\frac{d^2x}{ds^2} = -\frac{1-n}{\rho_0^2}x \tag{4.14}$$

The right-hand-terms in (4.9) and (4.13) express the focusing force. We rewrite (4.9) and (4.13) into a general form

$$\frac{d^2y}{ds^2} = -K(S)y \tag{4.15}$$

by defining a general focusing term $K(s)$ that has a periodicity $K(s + C) = K(s)$. Equation (4.15) is called Hill's equation.

Suppose that a solution of (4.15) has the form

$$y(s) = w(s)e^{i\psi(s)}. \tag{4.16}$$

Then (4.15) is divided into the two equations

$$w''(s) + K(s)w(s) - \frac{1}{w^3(s)} = 0 \tag{4.17}$$

and

$$\psi'(s) = \frac{1}{w^2(s)}. \tag{4.18}$$

According to Floque's theory the general solution of (4.15) can be expressed as

$$y(s) = c_1 w(s)e^{i\psi(s)} + c_2 w(s)e^{-i\psi(s)}. \tag{4.19}$$

Let the solutions be $y_1, w_1, \psi_1(y_2, w_2, \psi_2)$ at $s = s_1$ (s_2), respectively. Then the y_2 and y_2' are expressed by

$$\begin{pmatrix} y_2 \\ y_2' \end{pmatrix} = \begin{pmatrix} A & B \\ C & D \end{pmatrix} \begin{pmatrix} y_1 \\ y_1' \end{pmatrix}, \tag{4.20}$$

where

$$A = \frac{w_2}{w_1} \cos(\psi_2 - \psi_1) - w_2 w_1' \sin(\psi_2 - \psi_1)$$

$$B = w_1 w_2 \sin(\psi_2 - \psi_1)$$

$$C = -\frac{1 + w_1 w_1' w_2 w_2'}{w_1 w_2} \sin(\psi_2 - \psi_1) - \left(\frac{w_1'}{w_2} - \frac{w_2'}{w_1}\right) \cos(\psi_2 - \psi_1)$$

$$D = \frac{w_1}{w_2} \cos(\psi_2 - \psi_1) + w_1 w_2' \sin(\psi_2 - \psi_1).$$

This matrix is called a transfer matrix. This expression is similar to the ABCD matrix used in optics. Let us introduce the following parameters

$$\left.\begin{aligned}
\beta &:= w^2 \\
\alpha &:= -ww' \\
\gamma &:= \frac{1 + (ww')^2}{w^2}
\end{aligned}\right\} \tag{4.21}$$

$$\mu := \psi_2 - \psi_1. \tag{4.22}$$

The parameters β, α, γ are called Twiss parameters. The three Twiss parameters are not independent but have the relationship

$$\frac{1 + \alpha^2}{\beta} = \gamma. \tag{4.23}$$

By using (4.21) (4.20) is rewritten as

$$\begin{pmatrix} y_2 \\ y_2' \end{pmatrix} = \begin{pmatrix} \cos\mu + \alpha\sin\mu & \beta\sin\mu \\ -\gamma\sin\mu & \cos\mu - \alpha\sin\mu \end{pmatrix} \begin{pmatrix} y_1 \\ y_1' \end{pmatrix}. \tag{4.24}$$

Recalling that $y(s)$ is a linear combination of the two solutions (4.19) and $y(s)$ is a real number then we obtain

$$y(s) = cw(s) \cos(\psi(s) + \delta). \tag{4.25}$$

Differentiating (4.25),

$$y'(s) = \frac{c}{\sqrt{\beta}} \{-\alpha\cos(\psi(s) + \delta) - \sin(\psi(s) + \delta)\}. \tag{4.26}$$

Here we employ the following expressions

$$w'(s) = \frac{1}{2}\beta^{-1/2}\beta' = -\frac{\alpha}{\sqrt{\beta}}, \quad \psi'(s) = \frac{1}{\beta} \tag{4.27}$$

then (4.26) is written as

$$y'(s) = cw'(s)\cos(\psi(s) + \delta) - cw(s)\psi'(s)\sin(\psi(s) + \delta). \quad (4.28)$$

From (4.24) and (4.28) we express the cosine and sine functions using y and y'

$$\begin{cases} \cos\phi = \dfrac{y}{cw} \\[2ex] \sin\phi = -\alpha\cos\phi - \dfrac{\sqrt{\beta}y'}{c} \end{cases} \quad (4.29)$$

where $\phi = \psi(s) + \delta$. By removing ϕ we obtain

$$\gamma y^2 + 2\alpha yy' + \beta y'^2 = c^2. \quad (4.30)$$

This equation determines an ellipse in phase space $(x - x')$ with an area of πc^2. This area is independent from the particle position, thus is called Courant-Snyder invariant. Here we define an unnormalized emittance ε as phase space area divided by π, i.e. $\varepsilon := c^2$. Thus (4.30) is expressed as

$$\gamma y^2 + 2\alpha yy' + \beta y'^2 = \varepsilon, \quad (4.31)$$

and an beam ellipse of diverging beam ($\alpha < 0$) is depicted in Fig. 4.4. One can easily obtain that the maximum displacement as $x_{max} = \sqrt{\beta\varepsilon}$ and the maximum divergence angle as $x'_{max} = \sqrt{\gamma\varepsilon}$. The inclination angle ψ of the ellipse is

$$\tan 2\psi = -\frac{2\alpha}{\beta - \gamma}. \quad (4.32)$$

Recalling (4.25) and the definition of emittance we obtain

$$\begin{aligned} y(s) &= cw(s)\cos(\psi(s) + \delta) \\ &= \sqrt{\beta(s)\varepsilon}\cos(\psi(s) + \delta), \end{aligned} \quad (4.33)$$

which shows the amplitude of the betatron oscillation (transverse oscillation) to be $x_{max} = \sqrt{\beta\varepsilon}$.

Integrating (4.18) from s_1 to s_2 we obtain

$$\psi(s_2) - \psi(s_1) = \int_{s_1}^{s_2} \frac{1}{\beta(s)} ds. \quad (4.34)$$

The integration of the inverse of beta function gives the phase advance of the betatron oscillation.

Fig. 4.4 A beam ellipse drawn in the phase space $(x - x')$. This figure shows a diverging beam

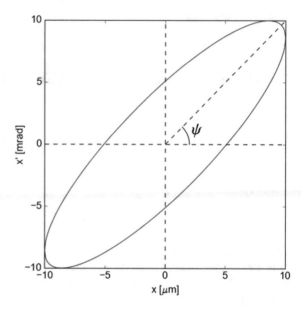

Comparison of laser beam Here let us compare the emittance and Twiss parameters of charged particles to lasers. A Gaussian beam of a laser has a beam parameter product

$$(BPP)_G := w_0 \theta_0 = \frac{\lambda}{\pi}, \tag{4.35}$$

where w_0 is the beam waist size at $1/e^2$ intensity and θ_0 is the half angle of the beam, and λ is a wavelength. Applying the same definition of beam emittance the emittance of a Gaussian laser beam can be

$$\varepsilon_L = (BPP)_G = \frac{\lambda}{\pi}. \tag{4.36}$$

Laser values are described in $1/e^2$ while charged particles are expressed in root-mean-square, thus the emittance for laser should be

$$\varepsilon_{L,rms} = \varepsilon_L/4 = \frac{\lambda}{4\pi}. \tag{4.37}$$

Practical laser beams are expressed in using M-square factor thus the practical rms emittance for laser beams is

$$\varepsilon_{L,rms,prac} = M^2 \frac{\lambda}{4\pi}. \tag{4.38}$$

For example, a Gaussian laser beam with a wavelength of $\lambda = 0.8\ \mu m$ has an unnormalized emittance of 0.064 mm-mrad. This corresponds to a normalized emittance of 10 mm-mrad for the electron beam with an energy of $E_e = 80$ MeV.

4.2.2 Longitudinal Beam Dynamics

Next let us consider motion of a charged particle under the influence of a sinusoidal electric field. The variable s stands for the synchronous particle that moves at the same phase velocity of the electric field. Then the equation of motion in the longitudinal space can be

$$\frac{d(W - W_s)}{ds} = qeE_0(\cos\phi - \cos\phi_s), \tag{4.39}$$

and

$$\frac{d(\phi - \phi_s)}{ds} = \frac{\omega}{c}\left(\frac{1}{\beta} - \frac{1}{\beta_s}\right) \approx -\frac{2\pi}{\lambda}\frac{W - W_s}{\gamma_s^3\beta_s^3 mc^2}. \tag{4.40}$$

In the final form, we expanded $1/\beta$ to the first order of γ and used the relationship $W = \gamma mc^2$. Differentiating (4.40) and using

$$\hat{s} = ks,\ \Delta\gamma = (W - W_s)/(mc^2),\ \Delta\phi = \phi - \phi_s \tag{4.41}$$

we obtain

$$\frac{\Delta\gamma}{d\hat{s}} = a_0(\cos\phi - \cos\phi_s), \tag{4.42}$$

and

$$\frac{\Delta\phi}{d\hat{s}} = -b_0(\Delta\gamma + \gamma_s). \tag{4.43}$$

Multiplying (4.42) by $d\delta/d\hat{s}$ and integrating by s we obtain

$$\frac{1}{2}(\Delta\gamma + \gamma_s)^2 = -\frac{a_0}{b_0}[\sin(\Delta\phi + \phi_s) - \Delta\phi\cos\phi_s] + C, \tag{4.44}$$

where C is a constant. This equation gives longitudinal phase space plots.

We can rewrite this form in Hamilton formalism:

$$H = -\frac{1}{2b_0}(\Delta\gamma + \gamma_s)^2 - a_0[\sin(\Delta\phi + \phi_s) - \Delta\phi\cos\phi_s], \tag{4.45}$$

and

$$\frac{d\Delta\phi}{d\hat{s}} = \frac{\partial H}{\partial\Delta\gamma},\ \frac{d\Delta\gamma}{d\hat{s}} = -\frac{\partial H}{\partial\Delta\phi}. \tag{4.46}$$

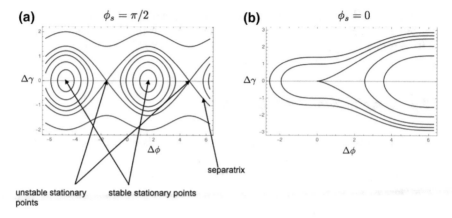

Fig. 4.5 Longitudinal phase space plots for $\phi_s = \pi/2$ (**a**) and $\phi_s = 0$ (**b**)

Example plots of longitudinal phase spaces are shown in Fig. 4.5. In this plot we can estimate the required acceptance, phase space area to achieve a certain beam performance.

4.2.3 Transfer Matrix

In this section we introduce transfer matrices to calculate the basic properties of beam optics. First, we modify the transverse motion by including an energy spread. Let us consider particles whose energy are different from the central energy. To do this we modify (4.1) by replacing $\gamma \to \gamma + \Delta\gamma$ and $v \to v + \Delta v$ and obtain

$$m\gamma \frac{d^2 x}{dt^2} = \frac{m(\gamma + \Delta\gamma)(v + \Delta v)^2}{\rho_0 + x} + qe(v + \Delta v)B_z(x). \qquad (4.47)$$

Taking only the first order terms and remembering that the forces are balanced at $x = 0$, i.e.,

$$\frac{m\gamma v^2}{\rho_0} + qev B_z(0) = 0, \qquad (4.48)$$

we obtain

$$\frac{d^2 x}{ds^2} + \frac{1-n}{\rho_0^2} x = \frac{1}{\rho_0} \frac{\Delta p}{p}. \qquad (4.49)$$

This is Hill's equation with a right-hand-side term. If $K = 1 - n/\rho_0^2$ is constant in a beam-optics element, we may have a special solution of (4.49) as

$$x = \frac{1}{\rho_0 K} \frac{\Delta p}{p}. \tag{4.50}$$

Thus a general solution of (4.49) can be written as (if $K > 0$)

$$x = C1 \cos(\sqrt{K}s) + C2 \sin(\sqrt{K}s) + \frac{1}{\rho_0 K} \frac{\Delta p}{p}. \tag{4.51}$$

Now we consider a new transfer matrix

$$\begin{pmatrix} x \\ x' \\ \Delta p/p \end{pmatrix} \tag{4.52}$$

instead of

$$\begin{pmatrix} x \\ x' \end{pmatrix}. \tag{4.53}$$

Setting x_0 and x_0' as the initial values at $s = 0$ the transfer matrix can be expressed as

$$\begin{pmatrix} x \\ x^{P}rime \\ \Delta p/p \end{pmatrix} = \begin{pmatrix} \cos(\sqrt{K}s) & \frac{1}{\sqrt{K}} \sin(\sqrt{K}s) & \frac{1}{\rho_0 K}\{1 - \cos(\sqrt{K}s)\} \\ -\sqrt{K} \sin(\sqrt{K}s) & \cos(\sqrt{K}s) & \frac{1}{\rho_0 K} \sin(\sqrt{K}s) \\ 0 & 0 & 1 \end{pmatrix} \begin{pmatrix} x_0 \\ x^{P}rime_0 \\ \Delta p/p \end{pmatrix} \tag{4.54}$$

for $K > 0$. Similarly we can get the case for $K < 0$ as

$$\begin{pmatrix} x \\ x' \\ \Delta p/p \end{pmatrix} = \begin{pmatrix} \cosh(\sqrt{|K|}s) & \frac{1}{\sqrt{|K|}} \sinh(\sqrt{|K|}s) & \frac{1}{\rho_0 - |K|}\{1 - \cosh(\sqrt{|K|}s)\} \\ -\sqrt{|K|} \sinh(\sqrt{|K|}s) & \cosh(\sqrt{|K|}s) & \frac{1}{\rho_0 \sqrt{|K|}} \sin(\sqrt{|K|}s) \\ 0 & 0 & 1 \end{pmatrix} \begin{pmatrix} x_0 \\ x_0' \\ \Delta p/p \end{pmatrix}. \tag{4.55}$$

6D transfer matrix To design matching sections between acceleration stages focusing is a key. Thus, we only consider beam transport components. If a beam line is composed of several optic elements, we can calculate beam parameters X_N after the optics using matrix calculations as $X_N = R_N R_{N-1} \cdots R_1 X_0$, where

$$R = \begin{bmatrix} R_{xx} & | & R_{xy} & | & R_{xz} \\ - - - & + & - - - & + & - - - \\ R_{yx} & | & R_{yy} & | & R_{yz} \\ - - - & + & - - - & + & - - - \\ R_{zx} & | & R_{zy} & | & R_{zz} \end{bmatrix} \tag{4.56}$$

is a six-dimensional matrix element and

$$X = \begin{bmatrix} x \\ x' \\ y \\ y' \\ z \\ \frac{\Delta p}{p} \end{bmatrix} \tag{4.57}$$

is a six dimensional beam parameter. Sometimes we do not need to calculate all six dimensions. One can specify what is needed in one's applications.

Twiss parameter calculation In this section we introduce Twiss parameters at a beam line optic element ($s = s_2$) from a point $s = s_1$. Here we consider two dimensions (x, x') and let a transfer matrix between s_2 and s_1 M,

$$M = \begin{pmatrix} m_{11} & m_{12} \\ m_{21} & m_{22} \end{pmatrix}. \tag{4.58}$$

From the definition of a transfer matrix (4.20) one can easily obtain

$$\begin{pmatrix} \alpha_2 \\ \beta_2 \\ \gamma_2 \end{pmatrix} = \begin{pmatrix} m_{11}m_{22} + m_{12}m_{21} & -m_{11}m_{21} & -m_{12}m_{22} \\ -2m_{11}m_{12} & m_{11}^2 & m_{12}^2 \\ -2m_{21}m_{22} & m_{21}^2 & m_{22}^2 \end{pmatrix} \begin{pmatrix} \alpha_1 \\ \beta_1 \\ \gamma_1 \end{pmatrix}. \tag{4.59}$$

For example, let us consider a drift space (no focusing and no acceleration) element. In this case $K \to 0$ and $\rho_0 \to \infty$ are inserted into (4.54) and we obtain

$$M_{drift} = \begin{pmatrix} 1 & s & 0 \\ 0 & 1 & 0 \\ 0 & 0 & 1 \end{pmatrix}. \tag{4.60}$$

To obtain this we used

$$\frac{\sin \sqrt{K}s}{\sqrt{K}} = s \frac{\sin \sqrt{K}s}{\sqrt{K}s} \xrightarrow[\sqrt{K}s \to 0]{} s. \tag{4.61}$$

The second example is a bending magnet with $n = 0$. In this case we take the following relationship

$$K_{Horiz} = \frac{1-n}{\rho_0^2} \to \frac{1}{\rho_0^2}, \ K_{Vert} = 0, \ L = \rho_0\theta \tag{4.62}$$

into (4.54) and we get

$$M_H = \begin{pmatrix} \cos\theta & \rho_0 \sin\theta & \rho_0(1 - \cos\theta) \\ -\frac{\sin\theta}{\rho_0}\cos\theta & \cos\theta & \sin\theta \\ 0 & 0 & 1 \end{pmatrix} \tag{4.63}$$

$$M_V = \begin{pmatrix} 1 & \rho_0\theta & 0 \\ 0 & 1 & 0 \\ 0 & 0 & 1 \end{pmatrix}. \tag{4.64}$$

The third example is a quadrupole (focusing device). We take the following relationship

$$K = \frac{G}{B\rho}, \rho_0 \to \infty \tag{4.65}$$

into (4.54) and we get

$$M_{QF} = \begin{pmatrix} \cos(\sqrt{K}s) & \frac{1}{\sqrt{K}}\sin(\sqrt{K}s) & 0 \\ -\sqrt{K}\sin(\sqrt{K}s) & \cos(\sqrt{K}s) & 0 \\ 0 & 0 & 1 \end{pmatrix} \tag{4.66}$$

and

$$M_{QD} = \begin{pmatrix} \cosh(\sqrt{-|K|}s) & \frac{1}{\sqrt{|K|}}\sinh(\sqrt{|K|}s) & 0 \\ -\sqrt{|K|}\sinh(\sqrt{K}s) & \cosh(\sqrt{|K|}s) & 0 \\ 0 & 0 & 1 \end{pmatrix}, \tag{4.67}$$

where $B\rho = pc/(qe)$ is called the *magnetic rigidity*.

Here, we show an application of the transfer matrix and calculate a beam parameter near the beam focus. Using the transfer matrix of a drift space with a length of L the beam parameter x can be expressed as

$$\begin{pmatrix} x \\ x' \end{pmatrix} = \begin{pmatrix} 1 & L \\ 0 & 1 \end{pmatrix} \tag{4.68}$$

$$\begin{pmatrix} x^* \\ x'^* \end{pmatrix}, \tag{4.69}$$

where x^* is the parameter at focus. Using (4.59) we obtain

$$\beta = -2L\alpha^* + \beta^* + L^2\gamma^*. \tag{4.70}$$

Using $\alpha^* = 0$ at focus and the relationship (4.23) i.e. $\gamma^* = 1/\beta^*$, (4.70) simplifies to

$$\beta(L) = \beta^* + \frac{L^2}{\beta^*}. \tag{4.71}$$

The beam spot size is $x(s) = \sqrt{\beta\varepsilon}$ and the beam size after focus is

$$x(L) = x^*\sqrt{1 + \left(\frac{L}{\beta^*}\right)^2}. \tag{4.72}$$

This expression is same as the Gaussian beam propagation

$$w(L) = w_0^* \sqrt{1 + \left(\frac{L}{Z_R}\right)^2}. \tag{4.73}$$

The beta function at focus β^* is the Rayleigh length of the charged particle beam.

4.3 Comparison of RF Accelerators and LWFA

In this section we discuss the field properties of radio-frequency accelerators and laser wakefield accelerators. What are similarity and difference between the two cases?

First we start with LWFA. In the linear regime of LWFA ($a_0 \ll 1$), longitudinal and transverse wake fields excited by a bi-Gaussian laser pulse ($a(r, \zeta) = a_0 \exp\left\{-r^2/(2\sigma_r^2) - \zeta^2/(2\sigma_z^2)\right\}$ are expressed by

$$eE_z(r, \zeta) = mc^2 \frac{\sqrt{\pi}}{2} k_p^2 \sigma_z a_0^2 \times \exp\left(-\frac{r^2}{2\sigma_r^2} - \frac{k_p^2 \sigma_z^2}{4}\right) \cos k_p \zeta, \tag{4.74}$$

$$eE_r(r, \zeta) = -mc^2 \sqrt{\pi} k_p \sigma_z a_0^2 \frac{r}{\sigma_r^2} \times \exp\left(-\frac{r^2}{2\sigma_r^2} - \frac{k_p^2 \sigma_z^2}{4}\right) \sin k_p \zeta, \tag{4.75}$$

respectively, where k_p is the plasma wavenumber, σ_r and σ_z are the transverse and longitudinal laser sizes. An example plot of the fields is given in Fig. 4.6a. Note that the phase difference of the longitudinal and transverse fields is π_4 and there exists a useful phase window where electrons are accelerated and focused. Note that the transverse field is dependent on the phase thus this causes an emittance growth. Such effects can be mitigated by using hollow plasma density profiles.

Then, let us see the situation in RF accelerators. Fields of a TM01 mode excited in a cylindrical cavity are given by

$$E_s = A J_0(kr) \exp i(\omega t - \beta_0 s), \tag{4.76}$$

$$E_r = i A \frac{\beta_0}{k} J_1(kr) \exp i(\omega t - \beta_0 s), \tag{4.77}$$

and

$$B_\theta = i A \frac{\omega}{kc^2} J_1(kr) \exp i(\omega t - \beta_0 s), \tag{4.78}$$

where J_n is the Bessel function of the first kind, $\beta_0 = 2\pi/\lambda_g$, and λ_g is the RF wavelength in a cavity, k is the wavenumber of the RF. The field distributions are

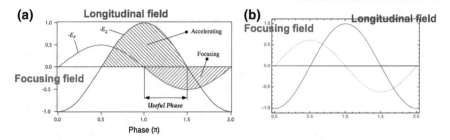

Fig. 4.6 Longitudinal and transverse focusing fields in LWFA (**a**) and a disk loaded RF accelerator (**b**)

depicted in Fig. 4.6b and are similar to those of LWFA. Thus, the longitudinal and focusing forces are expressed by

$$F_s = q A \cos(\omega t - \beta_0 s) \tag{4.79}$$

$$F_r = -\frac{q A \omega}{2} \frac{\omega}{c} \frac{1 - \beta_e \beta_w}{\beta/w} \cos(\omega t - \beta_0 s). \tag{4.80}$$

One of the differences between LWFA and RF accelerators can be that the focusing force for RF accelerators decreases as the beam energy becomes relativistic ($\beta_e \to 1$). This feature is favorable in terms of emittance conservation.

4.3.1 Liouvelle's Theorem and Emittance Conservation

In the last of the section we briefly give an important theorem *Liouville's theorem* in beam physics and comment on emittance conservation. Liouville's theorem states that the phase space density is conserved in a Hamiltonian system where the energy is conserved or there are no friction forces. In other words, the emittance cannot be improved in such system as focusing and external forces.

Even when Lioville's theorem is valid the actual emittance can be degraded by deforming the beam distribution is a complex structure as depicted in Fig. 4.7.

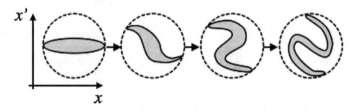

Fig. 4.7 An example of the beam distribution evolution in phase space

4.4 Review of Staged LWFA Experiments

In this section we review several experiments of staged laser acceleration experiments. Before starting we have to define what is the staging or staged acceleration. In LWFA two stages are just used to employ two stages which have two different plasma densities and are driven by the same laser pulse (2-stage, 1-laser). Only a few experiments employ two stages (two different plasma densities) and are driven by two independently-adjustable laser pulses (2-stage, 2-laser). An experimental setup of is depicted in Fig. 4.8. An ionization injection using a mixture gas (helium plus a small fraction of high-Z gas such as nitrogen) is commonly used and in the second stage a longer-length with standard single gas (e.g. pure helium) is used. We will discuss later but here we just stress the importance of the beam matching connecting to the stages. Beam matching is a kind of phase space manipulation so as to accelerate or transport the beam without loss or quality. In transverse matching a diverging beam, for example, from a stage should be properly focused onto the entrance of the second stage to meet the acceptance of the second stage. As well as transverse matching, longitudinal matching is also required.

Table 4.1 shows a summary of recent staged LWFA experiments [8–14]. All the experiments conducted so far except one used only a single laser and did not use special devices for beam matching (focusing), because of the complicated setup and difficulties in transporting and tuning the beams. In some experiments electron beams from the second stage had a similar amount of charge but usually the charges were lower than those from the 1st stages.

Evidence of two stage acceleration are (1) increase of energy compared to a single stage, (2) smaller beam divergence (mainly due to higher energy) than that obtained in the first stage, (3) energy gain dependence on the injection timing to the drive laser.

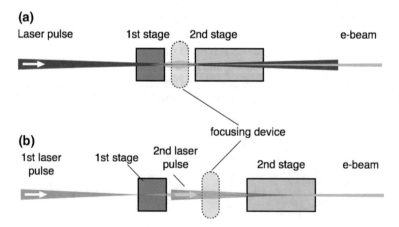

Fig. 4.8 A typical staged LWFA experiment setup. Two stages driven by a single laser pulse (**a**) and by two laser pulses (**b**)

Table 4.1 A typical staged LWFA experiment setup. Two stages driven by a single laser pulse (a) and by two laser pulses (b)

Group	Year	1st stage			Gap between stages/focusing scheme	2nd stage		Energy in 2nd stage
		Laser	Plasma	e-energy		Laser	Plasma	
UCLA	2011	1.7 J, 60 fs a_0=2–2.8	He:N_2 = 99.5%:0.5% L = 3 mm	<100 MeV	1 mm/no (probably plasma lens)	Same as 1st laser	He L = 5 mm	440 MeV 35 pC
SIOM	2011	40–60 TW, 40 fs	He:O_2 = 94%:6% L = 1 mm	<80 MeV	?/no (probably plasma lens)	Same as 1st laser	He L = 1 mm or 3 mm	Peaked at ~0.8 GeV (up to 1 GeV)
GIST	2013	25 J, 60 fs, w_0 = 25μm a_0 = 3.7	He L= 4 mm	0.4 GeV	~2 mm/ no (probably passive plasma lens)	Same as 1st laser	He L = 10 mm	Peaked at ~1 GeV (0.5–4 GeV?)
SIOM	2013	40–60 TW, 50 fs, w_0 = 16μm, f/20 a_0 = 2.0–2.5	He L = 0.8 mm (8–9) × 10^{18} cm^{-3} Down ramp	~30 MeV	-/no (probably passive plasma lens)	Same as 1st laser	He L = 1–5 mm, (2–6)×10^{18} cm^{-3}	310–530 MeV
Nebraska	2015	47 TW, 34 fs, w_{FH} = 20μm, f/14	He:N_2 = 99%:1% L = 0.5 mm 3.4 × 10^{18} cm^{-3}	<125 MeV	0.5 mm/passive plasma lens	Same as 1st laser	He L = 0.5 mm or 2 mm	Peaked at ~175 MeV (150–200 MeV)
LBNL	2016	1.3 J, 45 fs, f = 2 m w_0 = 18 μm a_0 = 1.4	He:N_2 = 99%:1% L = 0.7 mm 5 × 10^{18} cm^{-3}	120 MeV	55 cm/active plasma lens	0.45 J, 45 fs, f = 2 m w_0 = 18 μm a_0 = 0.81	H2 L = 33 mm capillary	Max. 270 MeV
SIOM	2016	100–120 TW, 33 fs w_{FH} = 32μm, f/30 a_0 = 1.3	He L = 0.8 mm 1.1 × 10^{19} cm^{-3}	Not shown	1–1.25 mm/(probably passive plasma lens)	Same as 1st laser	He L = 4 mm 6.0 × 10^{18} cm^{-3}	530–580 MeV 1% e-spread

In most of the experiments the beam matching between two stages are partly met by a so called passive plasma lens [15, 16], where the space-charge of an electron beam is weakened by background plasma. Steinke et al. employed an active plasma lens [17], which is capillary discharge device similar to a wave guiding device to increase an acceleration length in LWFA. In an active plasma lens, a discharge current of the order of hundreds amperes in the longitudinal direction can pinch an electron beam.

4.5 Beam Matching

Beam matching is used so that an input beam is matched to a lattice of the accelerator components so that the beam quality is conserved or kept as a design in the lattice. Applying this concept to LWFA is also possible and is very important to conserve the initial emittance and charge. Here, we show an example calculation using 2D particle-in-cell and tracking codes to estimate the acceptance of LWFA and the required beam parameters for that.

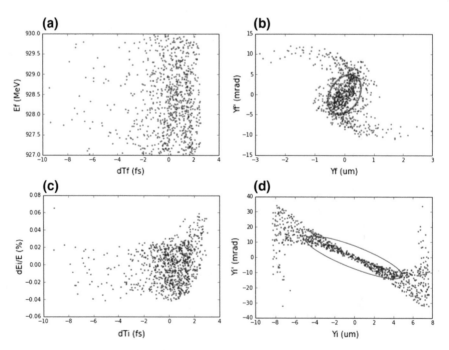

Fig. 4.9 Phase space plots of a PIC test particle calculation at the exit (**a** and **b**) and entrance (**c** and **d**). **a** and **c** show longitudinal phase space (but presentations are different in the two cases). **b** and **d** show the transverse phase space plots. The lines are fitted beam ellipses

The chosen parameters are a plasma density $n_e = 10^{-3}n_c$, a pulse duration $\tau = 30$ fs, a pulse energy $W_L = 5$ J, and a spot size $w_{FW} = 16\,\mu$m. The laser pulse excites plasma waves and non-interacting test particles are set to be accelerated in wakes. The initial test particles have a discrete energy of 20, 40, ..., 160 MeV and are wide enough both transversely and longitudinally to cover the plasma waves.

The calculation results are shown in Fig. 4.9. First, let us see the phase space plots at the exit (Fig. 4.9a). Because we have an interest to produce monochromatic electron beams, we select the electrons whose energies are within 0.3% centered at 938 MeV. This criteria is based upon one's interest. Such electrons have a beam divergence of a few mrad as seen in Fig. 4.9b and the normalized emittance is 5.9 mm-mrad. To produce such electrons we can get information how such electrons were distributed at the entrance of the LWFA as seen in Fig. 4.9c and d. Unfortunately, in this parameter regime the initial electrons should have a very monochromatic within $\Delta E/E = 0.11\%$ centered at 160 MeV (Fig. 4.9c). The initial beam should be a bit converging beam as seen in Fig. 4.9d, and the normalized emittance should be 14.6 mm-mrad. Beam matching can be made so as to satisfy one's requirement as we did in this section.

4.6 Comparison of Focusing Devices

In this section we describe focusing devices to achieve transverse beam matching in staged laser wakefield acceleration. A brief summary of the comparison is given in Table 4.2.

In conventional accelerators electromagnet quadrupoles are commonly used. A quadrupole magnet has a focusing force in one direction (e.g. horizontal), while the perpendicular direction (vertical) has a defocusing force. Thus a doublet of

Table 4.2 A comparison of focusing devices. Simplified formulas are given for quick comparison. Also, example focal lengths are calculated for 1 GeV electron beams with moderate parameters

	L	B or B'	Symmetric	Focal length (thin lens)	Focal length for 1GeV e-
Quadrupole	~20 cm	10–100 T/m	No	$f = \frac{B\rho}{B'L}$	>0.167 m
Triplet of Q	~10+20+10	10–100 T/m	Yes	$f = \frac{6(B\rho)^2}{B'^2L^3} \quad L/2 + L + L/2$	>0.834 m
Solenoidal lens	~10 cm	2 T	Yes	$f = \frac{(2B\rho)^2}{B^2L}$	55 m
Passive plasma lens	1 mm	–	Yes	$f = \frac{2\varepsilon_0(B\rho)^2}{Lmn} = \frac{\gamma}{2\pi r_e nL}$	335 μm (nb*[1] ~ 3.3 × 10^{17} cm^{-3})
Active plasma lens	~10 cm	3000 T/m	Yes	$f = \frac{B\rho}{B'L}$	11 mm

quadrupole magnets is widely used to have a focusing force for two directions despite the astigmatic feature, which means the two focal positions are different. A symmetric lens can be achieved with a triplet of quadrupole magnets. The focal length of a triplet is

$$f = \frac{6(B\rho)^2}{B'^2 L^3},$$ (4.81)

where the separation distances are zero and the lengths of the quadrupoles are $L/2, L, L/2$, respectively. The triplet has a symmetric focusing property while the focal length gets longer.

Instead of electromagnets, permanent magnets can be used. Recently variable magnetic fields are available with permanent-magnet based dipole and quadrupoles [18] by moving some parts that form magnetic circuits. Variable permanent magnet quadrupoles are beneficial because of compactness, strength of fields, and unnecessity of water-cooling. Thus, this device has potentially good matching to LWFA.

A symmetric lens can be achieved by using a solenoidal magnet. However, the focal length of a solenoidal magnet is proportional to the square of the beam momentum and the inverse of the magnet length. This is not effective for high energy particles. Therefore, it is used in low energy sections. A solenoid magnet can be also fabricated with permanent magnets [19]. Electromagnets use direct current (DC) but can use larger current based on pulse power technology to increase the power of focusing. Such magnets are used in Japan [20].

The above mentioned technologies are widely used in conventional accelerators and, here, we introduce plasma based focusing devices. First, we show a passive plasma lens, where the space-charge of the charged particle beams is partially or fully compensated by the plasma and thus the remaining self-B field can focus the beams [15, 16]. In this regime, beam-driven wakes are sometimes excited and the wake field can work as a focusing device as well. As mentioned in the previous section, almost all LWFA experiments include this effect regardless of intention or not, because the gas extends towards peripheral vacuum regions and the passing drive laser is intense enough to ionize the gas. This effect was clearly observed in a separate experiment [21].

The second device is an active plasma lens, where a discharge current in a gas-filled capillary produces strong focusing forces as

$$f = \frac{B\rho}{B'L},$$ (4.82)

where $B' = \mu_0 I_0/(2\pi R^2)$, μ_0 is the vacuum permeability, I_0 is the peak current, R is the capillary radius [17]. As shown in Table 4.2, the focal length of an active plasma lens can be very short and has an axial symmetry.

As we explained, transverse focusing is a necessary device in future, realistic machines based on staged LWFA. LWFA itself can provide a compact accelerators but the benefit would be degraded if we have to use standard (electromagnets). Thus, the above mentioned plasma devices or new concepts will be also necessary.

4.7   Further Topics

In this section we point out important topics that are not discussed in this lecture.

Space-charge effects

Because the LWFA can produce a dense beam in a 6D phase space, the space-charge effects should be properly taken into account especially in low energy beam transports.

Beam loading effects

In Sect. 4.5 we show an example to estimate the beam matching in a LWFA stage. However, we did not take beam loading effects into the calculation because of the simplicity. Test particle method is useful to estimate the overall beam qualities at first. Then, the final adjustment are needed using several iterations of self-consistent calculations using a PIC code, defining a size, duration, and charge of the initial electron beam. The beam loading may modifies a structure of wake fields and in this sense, the employment of a PIC code is inevitable especially for a high-charge case.

Longitudinal beam matching

Most of the efforts in this paper is focused on the transverse beam matching. However, the longitudinal beam matching is also important as well. In particular, an electron beam with relatively large energy spread is elongated in time even when it propagates in vacuum because of the velocity difference. In addition, the space-charge force accelerates the head part of the beam and decelerates the tail part. To prevent this, a dog-leg or an alpha magnet can be used to compress the electron bunch in order to match the acceptance of the next stage. In other methods we can choose proper accelerating phase to compensate the energy difference to get a monochromatic energy.

Coherent synchrotron radiation

In a bunch compression system, a short bunch emits the coherent synchrotron radiation (CSR) provided that the bunch length is shorter than the emitted radiation wavelength. Of course this causes the energy loss of the beam, but what is worse is that the CSR can affect the bunch itself because of a curved orbit in magnets. This cause a chirped distribution of the electron bunch and degrades the beam quality. CSR effects can be mitigated by double bend achromat optics [22].

4.8   Conclusion

In this lecture we provide the fundamentals of beam physics, a recent review of staged laser wakefield acceleration (LWFA), an example calculation of the acceptance of a LWFA stage, and a brief comparison of focusing devices. Staged LWFA is straightforward progress if we look into standard linear accelerators to achieve higher energy, or control beam quality. The progress of the LWFA theory and experimental techniques proved that a single stage LWFA can produce high energy, ultra-short, low emittance beams. The remaining tasks are to improve overall stabilities, an improvement on energy spread. We need to further improve the qualities and probably several technological and fundamental breakthroughs would be necessary.

Acknowledgements We thank Dr. T. Esirkepov for providing PIC simulation data, Dr. J. Koga for careful reading of the manuscript, Dr. T. Hosokai, Dr. K. Huang, Dr. M. Mori, Dr. N. Nakanii, Dr. Y. Hayashi, Dr. H. Kotaki, Prof. S. V. Bulanov for fruitful discussions, and Dr. Y. Sano for encouragement. This work was supported by the ImPACT Program of the Council for Science, Technology and Innovation (Cabinet Office, Government of Japan).

References

1. T. Hara et al., Pulse-by-pulse multi-beam-line operation for x-ray free-electron lasers. Phys. Rev. Accel. Beams **19**, 020703 (2016)
2. Japan Science and Technology Agency. http://www.jst.go.jp/impact/en/program/03.html
3. P.A. Walker, Horizon 2020 EuPRAXIA design study, in *Proceedings of International Particle Accelerator Conference*, vol. 2017 (2017), p. 1265
4. E.D. Courant, H.S. Snyder, Theory of the alternating-gradient synchrotron. Ann. Phys. **3**, 1–48 (1958)
5. CERN Accelerator School. http://cas.web.cern.ch
6. U.S. Particle Accelerator School. http://uspas.fnal.gov
7. KEK Accelerator School (OHO) [in Japanese]. http://accwww2.kek.jp/oho/index.htm
8. B.B. Pollock et al., Demonstration of a narrow energy spread ~ 0.5 GeV electron beam from a two-stage laser wakefield accelerator. Phys. Rev. Lett. **107**, 1191 (2011)
9. J.S. Liu et al., All-optical cascaded laser wakefield accelerator using ionization-induced injection. Phys. Rev. Lett. **107**, 035001 (2011)
10. H.T. Kim et al., Enhancement of electron energy to the multi-GeV regime by a dual-stage laser-wakefield accelerator pumped by petawatt laser pulses. Phys. Rev. Lett. **111**, 165002 (2013)
11. W. Wang et al., Control of seeding phase for a cascaded laser wakefield accelerator with gradient injection. Appl. Phys. Lett. **103**, 243501 (2013)
12. G. Golovin et al., Tunable monoenergetic electron beams from independently controllable laser-wakefield acceleration and injection. Phys. Rev. Spec. Top. Accel. Beams **18**, 011301 (2015)
13. S. Steinke et al., Multistage coupling of independent laser-plasma accelerators. Nature **530**, 190–193 (2016)
14. W.T. Wang et al., High-brightness high-energy electron beams from a laser wakefield accelerator via energy chirp control. Phys. Rev. Lett. **117**, 1988 (2016)
15. P. Chen, Plasma focusing and diagnosis of high energy particle beams. SLAC-PUB-5186 (1990)
16. J.J. Su, T. Katsouleas, J.M. Dawson, R. Fedele, Plasma lenses for focusing particle beams. Phys. Rev. A **41**, 3321–3331 (1990)
17. J. van Tilborg et al., Active plasma lensing for relativistic laser-plasma-accelerated electron beams. Phys. Rev. Lett. **115**, 184802 (2015)
18. F. Marteau et al., Variable high gradient permanent magnet quadrupole (QUAPEVA). Appl. Phys. Lett. **111**, 253503 (2017)

19. M. Kando et al., Matching section to the RFQ using permanent magnet symmetric lens, in *Proceedings of the 1995 Particle Accelerator Conference*, vol. 1843 (1995)
20. Y. Sakai et al., Narrow band and energy selectable plasma cathode for multistage laser wakefield acceleration. Phys. Rev. Accel. Beams **21**, 101301 (2018)
21. S. Kuschel et al., Demonstration of passive plasma lensing of a laser wakefield accelerated electron bunch. Phys. Rev. Accel. Beams **19**, 171 (2016)
22. T. Hara et al., High peak current operation of x-ray free-electron laser multiple beam lines by suppressing coherent synchrotron radiation effects. Phys. Rev. Accel. Beams **21**, 040701 (2018)

Chapter 5
Fundamentals and Applications of Hybrid LWFA-PWFA

Bernhard Hidding, Andrew Beaton, Lewis Boulton, Sebastién Corde,
Andreas Doepp, Fahim Ahmad Habib, Thomas Heinemann, Arie Irman,
Stefan Karsch, Gavin Kirwan, Alexander Knetsch, Grace Gloria Manahan,
Alberto Martinez de la Ossa, Alastair Nutter, Paul Scherkl, Ulrich Schramm
and Daniel Ullmann

Abstract Fundamental similarities and differences between laser-driven plasma wakefield acceleration (LWFA) and particle-driven plasma wakefield acceleration (PWFA) are discussed. The complementary features enable the conception and development of novel hybrid plasma accelerators, which allow previously not accessible compact solutions for high quality electron bunch generation and arising applications. Very high energy gains can be realized by electron beam drivers even in single stages because PWFA is practically dephasing-free and not diffraction-limited. These electron driver beams for PWFA in turn can be produced in compact LWFA stages. In various hybrid approaches, these PWFA systems can be spiked with ionizing laser pulses to realize tunable and high-quality electron sources via optical

This paper is a reprint of the paper in Appl. Sci. 2019, 9(13), 2626; https://www.mdpi.com/2076-3417/9/13/2626/htm.

B. Hidding (✉) · A. Beaton · L. Boulton · F. A. Habib · T. Heinemann · G. Kirwan ·
G. G. Manahan · A. Nutter · P. Scherkl · D. Ullmann
Department of Physics, University of Strathclyde, 107 Rottenrow, Glasgow G40NG, UK
e-mail: bernhard.hidding@strath.ac.uk

A. Beaton
e-mail: andrew.beaton@strath.ac.uk

L. Boulton
e-mail: lewis.boulton@strath.ac.uk

F. A. Habib
e-mail: ahmad.habib@strath.ac.uk

T. Heinemann
e-mail: thomas.heinemann@strath.ac.uk

G. Kirwan
e-mail: gavin.kirwan@strath.ac.uk

G. G. Manahan
e-mail: grace.manahan@strath.ac.uk

L. A. Gizzi et al. (eds.), *Laser-Driven Sources of High Energy Particles and Radiation*,
Springer Proceedings in Physics 231, https://doi.org/10.1007/978-3-030-25850-4_5

density downramp injection (also known as plasma torch) or plasma photocathodes (also known as Trojan Horse) and via wakefield-induced injection (also known as WII). These hybrids can act as beam energy, brightness and quality transformers, and partially have built-in stabilizing features. They thus offer compact pathways towards beams with unprecedented emittance and brightness, which may have transformative impact for light sources and photon science applications. Furthermore, they allow the study of PWFA-specific challenges in compact setups in addition to large linac-based facilities, such as fundamental beam–plasma interaction physics, to develop novel diagnostics, and to develop contributions such as ultralow emittance test beams or other building blocks and schemes which support future plasma-based collider concepts.

5.1　Introduction and Fundamental Considerations

Laser-driven wakefield accelerators (LWFA) and electron-driven plasma wakefield accelerators (PWFA) rely on similar concepts [1]. An intense laser or electron beam driver propagates through a plasma, and, based on its typically mainly transverse forces, plasma electrons are kicked out of the path of the driver beam, not unlike how a snowplough operates. However, unlike with a snowplough, these electrons then feel the re-attractive forces of the plasma ions, which have four orders of magnitude more mass than the electrons and therefore can be assumed quasistatic in first approximation. This transient charge separation follows the driver beam through the plasma and generates a co-moving plasma oscillation wave with more or less spherical geometry. The length of the arising bubble (LWFA) or blowout (PWFA)

A. Nutter
e-mail: alastair.nutter@strath.ac.uk

P. Scherkl
e-mail: paul.scherkl@strath.ac.uk

D. Ullmann
e-mail: daniel.ullmann@strath.ac.uk

B. Hidding · A. Beaton · L. Boulton · F. A. Habib · T. Heinemann · G. Kirwan · G. G. Manahan · A. Nutter · P. Scherkl · D. Ullmann
Cockcroft Institute, Sci-Tech Daresbury, Keckwick Lane, Daresbury, Cheshire WA4 4AD, UK

L. Boulton · T. Heinemann · G. Kirwan · A. Knetsch · A. de la Ossa
Deutsches Elektronen-Synchrotron DESY, 22607 Hamburg, Germany
e-mail: alexander.knetsch@desy.de

A. de la Ossa
e-mail: alberto.martinez.de.la.ossa@desy.de

S. Corde
LOA, ENSTA ParisTech, CNRS, Ecole Polytechnique, Université Paris-Saclay, 91762 Palaiseau, France
e-mail: sebastien.corde@polytechnique.edu

structure is determined by the plasma electron density n_e and the plasma wavelength $\lambda_p = 2\pi c[\epsilon_0 m_e/(e^2 n_e)]^{1/2}$, which is solely dependent on n_e and where c is the speed of light, ϵ_0 is the vacuum permittivity, and m_e is the electron mass and e the elementary charge. In the ideal case, the driver beam length σ_z or laser pulse duration τ fits into half of the plasma wavelength such that $c\tau \approx \lambda_p/2$, because then, as with a harmonic oscillator such as a pendulum, the excitation is strongest because the driver excites the plasma wave just until the zero-crossing of the oscillatory movement. In terms of the plasma wave number $k_p = 2\pi/\lambda_p$, this desirable relation between driver length σ_z and plasma wave number k_p can be expressed as $k_p \sigma_z < 1$.

Figure 5.1 visualizes (based on particle-in-cell simulations with the code VSim [2]) how the driver (green) excites a plasma blowout/bubble with large accelerating and decelerating fields up to tens of GV/m (color-coded).

Here, the plasma density is $n_p = 10^{17}\,\text{cm}^{-3}$, the corresponding plasma wavelength $\lambda_p \approx 106\,\mu\text{m}$, and the plasma skin depth $k_p^{-1} \approx 16\,\mu\text{m}$. The width of the driver pulse is in both cases set to $\sigma_r = 10\,\mu\text{m}$. The strength of the driver is characterized by the dimensionless electron beam charge $\tilde{Q}$ and the dimensionless laser light amplitude a_0, respectively. Here, the dimensionless light amplitude, defined by the ratio of electric field E and frequency ω, amounts to $a_0 = \frac{eE}{m_0\omega c} \approx 3$. The dimensionless beam charge [3], which is defined by the ratio of the beam electron density per cubic skin depth $N_b k_p^3$ and the background plasma density $\tilde{Q} = \frac{N_b k_p^3}{n_p} = 4\pi k_p r_e N_b$, where $r_e = 4\pi\epsilon_0 c^2 m$ is the classical electron radius, which is also set to $\tilde{Q} \approx 3$, since the values of a_0 and $\tilde{Q}$ define the interaction regime of LWFA and PWFA, respectively (see Fig. 5.2).

Notably, the plasma density is not relevant for the type of LWFA interaction, just the laser frequency and intensity, whereas, for PWFA, $\tilde{Q} \propto \sqrt{n_e}$. The background of this is that a particle beam has unipolar fields, while a quickly oscillating laser pulse can penetrate plasma up to the critical density $n_c(\omega) = \epsilon_0 m_0 \omega^2/e^2$, which amounts to $n_c \approx 1.7 \times 10^{21}\,\text{cm}^{-3}$ for a typical Ti:sapphire laser pulse.

A. Doepp · S. Karsch
Ludwig-Maximilians-Universität München, Am Coulombwall 1, 85748 Garching, Germany
e-mail: andreas.doepp@mpq.mpg.de

S. Karsch
e-mail: stefan.karsch@mpq.mpg.de

Helmholtz-Zentrum Dresden-Rossendorf, Institute of Radiation Physics, Bautzner Landstrasse 400, 01328 Dresden, Germany

A. Irman · A. Nutter · U. Schramm
Max Planck Institut ü Quantenoptik, Hans-Kopfermann-Str. 1, 85748 Garching, Germany
e-mail: a.irman@hzdr.de

U. Schramm
e-mail: u.schramm@hzdr.de

D. Ullmann
Central Laser Facility, STFC Rutherford Appleton Laboratory, Didcot, Oxfordshire OX11 0QX, UK

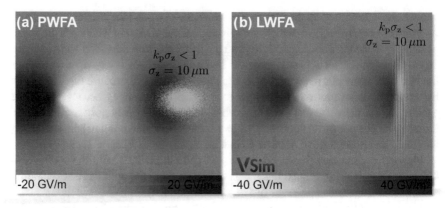

Fig. 5.1 PIC-simulation visualization of electron-driven PWFA (**a**) and laser-driven LWFA (**b**) in the blowout/bubble regime, respectively. The driver beam is shown in green and propagates to the right, expels plasma electrons (not shown) and thus generates strong trailing electric decelerating/accelerating fields

(a)

$$\tilde{Q} = \frac{N_b k_p^3}{n_p} \begin{cases} < 1 & \text{linear regime} \\ \approx 1 & \text{quasilinear} \\ > 1 & \text{nonlinear (blowout regime)} \end{cases}$$

(b)

$$a_0 = \frac{eE}{m_0 \omega c} \begin{cases} < 1 & \text{linear regime} \\ \approx 1 & \text{quasilinear} \\ > 1 & \text{nonlinear (bubble regime)} \end{cases}$$

Fig. 5.2 Linear, quasi-linear and non-linear regimes of PWFA (**a**) and LWFA (**b**), as indicated by the corresponding interaction value of $\tilde{Q}$ and a_0, respectively

To excite the longitudinally trailing plasma wave, it is required to push plasma electrons mainly in the transverse direction as efficiently as possible. While a point charge (or an ensemble of point charges) has a radial Coulomb electric field distribution $E_r = \frac{e}{4\pi\epsilon_0} \frac{1}{r^2}$, a Lorentz transformation of the electric fields yields $E'_\perp = \gamma E_r \gg E'_\parallel = E_r$, such that the electric field (more accurately $E'_\perp = \gamma(\mathbf{E} + \mathbf{v} \times \mathbf{B})$) of a relativistically moving electron is mainly transversal, scaling with the Lorentz factor $\gamma = (1 - v^2/c^2)^{-1/2}$. For the magnetic field $B_0 = E_0/c$, we have $B'_\parallel = 0$ and $B'_\perp = vE_r c^{-2}$. The case of an electron bunch interacting with plasma is much more complex, as magnetic self-fields, return currents, etc. have to be taken into account.

Figure 5.3 visualizes the transverse field of an electron driver in the lab frame. The Coulomb force $F_{\text{Coulomb}} = e\mathbf{E}$ of an electron bunch on plasma electrons is unipolar and scales linear with the perceived electric field. This is highly suitable in order to expel electrons off axis and to set up a plasma wave.

This is an important difference to laser drivers, where the electric field is not unipolar, but oscillates and puts the plasma electrons into a quiver motion. Hence, the plasma electron expulsion of electrons is achieved by means of the pondermotive force $F_{\text{pond}} = -\frac{e^2}{4m\omega^2} \nabla \mathbf{E}^2 \propto \nabla I$, which scales with the gradient of the electric field squared or intensity I. As a result, the electric fields which produce a similarly strong plasma wake excitation are orders of magnitude lower for an electron bunch

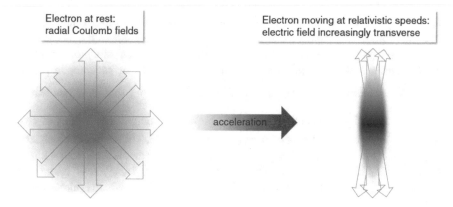

Fig. 5.3 The electric field of a point charge in the lab frame is mainly transverse, scaling with its Lorentz factor γ

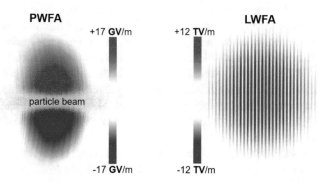

Fig. 5.4 The electric fields of electron bunch PWFA drivers are unipolar, while electric fields of laser pulse LWFA drivers are oscillating. As a result, the peak electric fields, which are required to excite a similarly strong blowout/bubble, are three orders of magnitude lower for PWFA than for typical LWFA Ti:sapphire lasers

PWFA driver when compared to a laser pulse LWFA driver, $E_{r,\text{PWFA}} \ll E_{r,\text{LWFA}}$. This fundamental difference is visualized in Fig. 5.4, showing that the electric fields of an electron bunch driver as used for Fig. 5.1 are three orders of magnitude lower than for a correspondingly strong laser driver as used for Fig. 5.1. This has important implications, in particular for the choice of media and injection mechanisms.

Now, in a bunch of particles instead of only one electron, space charge forces will make the bunch diverge, and lengthen. However, as regards lengthening or "dephasing" ΔL of a bunch over an acceleration distance L of a sufficiently relativistic and monoenergetic beam $[\gamma; \gamma + \Delta\gamma]$ such that $\Delta\gamma \ll \gamma$, the lengthening $\frac{\Delta L}{L} \approx \frac{1}{\gamma^2}\frac{\Delta\gamma}{\gamma}$ is typically very small. While bunch lengthening is typically negligible even over meter-scale distances in vacuum, energy transfer and thus deceleration of a drive beam in plasma leads to increasing ΔL and therefore to significant lengthening, up to the point where drive beam electrons can slow down so much that they fall back and travel from the decelerating into the accelerating phase of the blowout [4].

For laser pulses drivers, we have a fundamentally different situation. While an electromagnetic wave propagates with the speed of light in vacuum, in plasma the group velocity of a laser pulse is $v_g = c \left(1 - \frac{\omega_p^2}{\omega_0^2}\right)^{1/2}$, thus is dependent on the plasma density n_e via $\omega_p = \left(\frac{n_e e^2}{m_e \epsilon_0}\right)^{1/2}$. Figure 5.5 compares velocities of electrons and laser pulses in terms of corresponding $\beta = v/c$. The energy of an electron $E_{\text{kin}} = m_0 c^2 (\gamma - 1) \approx 0.511\,\text{MeV}\,(\gamma - 1)$, and, e.g., for an electron with a kinetic energy of 0.5 MeV, the corresponding $\gamma \approx 1.97$ and $\beta \approx 0.54$, whereas, for an electron at 5 MeV, its Lorentz factor is $\gamma \approx 10.87$ and $\beta \approx 0.9957$, which is already close to the speed of light.

This has an importance consequence, namely that, in LWFA, accelerated electrons will outrun the laser driver. They therefore move from the accelerating field phase in the back of the bubble forward into the decelerating field in the first part of the bubble, which limits the useful acceleration distance and hence energy gain. It is interesting to note that, as pointed out above, in LWFA, the wake excitation strength is independent of the plasma density, but the acceleration distance is, whereas in contrast in PWFA the wake excitation strength is dependent on the plasma density, but the acceleration distance is not. This is why PWFA, a sufficiently intense and energetic driver provided, can harness phase-constant, tens of GV/m-scale (or higher) accelerating fields over meter-scale distance, and hence realize tens of GeV energy gains in a single stage.

Another important aspect that arises from the different wake velocities in LWFA and PWFA regards injection and dark current, respectively. While in PWFA, due to practically negligible dephasing, the wake velocity equals the driver velocity such that $\gamma_{\text{wake}} \approx \gamma_p \approx 10^4$ for a 10 GeV driver beam, the wake velocity of the plasma wave is that of the driving laser beam such that $\gamma_{\text{wake}} \approx \gamma_p = \left(1 - \frac{v_p^2}{c^2}\right)^{-1/2} \approx 10 - 100$ for typical plasma densities. A consequence of this is that electron injection and bunch generation via self-injection or other mechanisms is considerably easier to achieve in LWFA than in PWFA, while in turn it is easier to keep PWFA dark-current-free by excluding unwanted self-injection mechanisms. The general trapping condition for injected electrons is that the injected electrons have to catch up with the plasma wave such that $\gamma_{\text{electron}} \geq \gamma_{\text{wake}} \approx \gamma_p$. The Hamiltonian is the sum of a kinetic energy term and the potential energy. If injected electrons are assumed to be initially at rest, this means the electrostatic wakefield potential energy in the wave frame needs to be larger than the required electron kinetic energy $-e\Delta\phi' \geq E'_{\text{kin}} = E_0(\gamma_p - 1) = m_0 c^2 (\gamma_p - 1)$. The reduced threshold for injection due to lower γ_p in LWFA is helpful if aiming at producing electron bunches on the fly, but makes it more difficult to avoid dark current, and in turn the higher threshold for injection makes it is more difficult to produce electron bunches on the fly in PWFA, but on the other hand is advantageous in order to realize a robust, dark-current free system.

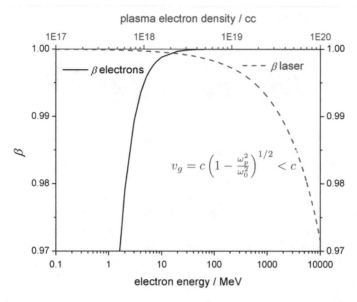

Fig. 5.5 The velocity of electrons in plasma is independent of plasma density, while the laser pulse group velocity in plasma decreases with increasing plasma density. Electron velocities in terms of corresponding β are plotted in black versus electron energy, while the laser pulse group velocity is plotted in red versus plasma density

As regards divergence and diffraction, both electron beams and the laser pulses expand hyperbolically when propagating in the forward direction z. A Gaussian laser pulse focused to a spot size w_0 exhibits diffraction following $w(z) = w_0\sqrt{1 + (z/Z_R)^2}$. Similarly, for an electron beam focused to a spot size σ_{r0}, the transverse evolution is $\sigma_r = \sigma_{r0}\sqrt{1 + (z/\beta^\star)^2}$. However, while the Rayleigh length $Z_R = \pi w_0^2/\lambda$ of a typical high-power Ti:sapphire laser pulse with central wavelength $\lambda \approx 0.8\,\mu$m is tunable only via the spot size, which in turn needs to be small in order to yield high intensity for a given laser power, the beta function $\beta^* = \sigma_{r0}^2\gamma/\epsilon_n$ of an electron pulse can not only be tuned via the spot size σ_{r0}, but also via the electron γ and normalized emittance ϵ_n. More importantly, it is also much larger for typical values of γ and normalized emittance ϵ_n. For example, for the same spot size of laser and electron beam $w_0 = \sigma_{r0} = 10\,\mu$m, the Rayleigh range is $Z_R \approx 400\,\mu$m for Ti:sapphire pulses, while $\beta^* \approx 20$ cm for a typical electron beam energy of 1 GeV, which corresponds to $\gamma \approx 2000$, and a typical normalized emittance of the drive beam of $\epsilon_n = 10^{-6}$ mrad. This dramatically different length over which an electron driver beam and laser pulse, respectively, stay compact, is sketched in Fig. 5.6. This is another main reason PWFA can straightforwardly realize substantially longer acceleration distances, and hence energy gains in a single stage, than LWFA.

It should be noted that self-focusing or active guiding, e.g., in a transversely parabolic density channel, can extend the effective Rayleigh range and hence the obtainable diffraction-limited acceleration distance of an LWFA driver pulse

PWFA   **LWFA**

Fig. 5.6 An electron beam of typical energy and emittance stays focussed over much longer length than a typical laser pulse, focused to the same spot size

substantially. For an electron beam driver, there is a contribution to the divergence due to intra-bunch space charge forces. The radial Lorentz force of a particle inside the bunch is $F_r = e(E_r + v \times B_\theta)$ and the electric field within an (infinitely) long bunch with uniform density n_b is given by Gauss' law as $E_r = \frac{1}{2} e n_b r / \epsilon_0$ and the magnetic field by Faraday's law as $B_\theta = \frac{1}{2} e n_b \mu_0 r$. With $c^2 = 1/(\epsilon_0 \mu_0)$, this yields $F_r = e E_r \left(1 - \frac{v^2}{c^2}\right) = e E_r \gamma^{-2}$. This means that transverse forces are limited, and also that for a witness bunch which is produced in a plasma wave (both driven by lasers as well as by particle beams) the transverse space charge forces quickly decrease during acceleration as γ^{-2}. For a highly relativistic ($\gamma \gg 1$) bunch propagating in vacuum, the current-induced pinching magnetic field balances the Coulomb space charge repulsion.

However, when propagating in plasma, the repulsive electron bunch space charge forces can be shielded by the plasma electrons, such that in total a net focusing force arises due to the magnetic field. To some extent, this "plasma lens" is the equivalent of relativistic self-focusing in LWFA. In some analogy to the classification of laser propagation through plasma, one differentiates between an underdense plasma lens, where the beam density $n_b \gg n_p$, and an overdense plasma lens where the plasma density is so high that $n_b \ll n_p$. The underdense plasma lens, a $\tilde{Q} > 1$ case in the non-linear interaction regime, is straightforward because it simply means that the plasma electrons are expelled from the vicinity of the electron beam and thus a uniform ion background shields the space charge of the beam, and the focusing strength scales linearly with ambient plasma density n_p in principle without spherical aberrations. The head of the bunch, and depending on plasma skin depth and width and length of the bunch other parts of the bunch may not be seeing a fully uniform pure ion background. In the overdense plasma lens, the electron bunch space charge is only a small perturbation and the collective shielding by plasma electrons takes place locally and is proportional to the local bunch density. For example, in the case of a typical Gaussian electron bunch with density distribution $n_b(x, y, z) = n_{b0} \exp \frac{-x^2}{2\sigma_x^2} \exp \frac{-y^2}{2\sigma_y^2} \exp \frac{-z^2}{2\sigma_z^2}$, the peak bunch density in the center $\hat{n}_{b,max} = N/((2\pi)^{3/2} \sigma_r^2 \sigma_z)$ and thus the focusing strength is much larger than farther outside, which results in spherical aberrations. It helps if the bunch is long compared to the skin depth of the plasma such that $k_p \sigma_z \gg 1$, and if the beam is narrow such that $k_p \sigma_r \ll 1$.

In summary, the above considerations reveal a high degree of complementarity of LWFA and PWFA. Main features and considerations of employing laser pulses and electron beam pulses for plasma wakefield acceleration are:

- Electron bunches drive plasma waves efficiently due to unidirectional fields, already at comparably low bunch self-field values such as few GV/m to tens of GV/m.
- Laser pulses require large electric fields, typically in the TV/m range, to excite plasma waves via the ponderomotive force due to their oscillating field structure.
- Laser pulses are very efficient to tunnel ionize matter and hence to provide plasma, due to the high peak electric laser pulse fields.
- Electron bunches are not efficient for ionizing matter because of the low electric self-fields.
- Electron bunches propagate with approximately the speed of light even in plasma, whereas dephasing is a fundamental problem in LWFA. This allows much longer, and phase constant acceleration to be realized in PWFA.
- Electron bunches are stiff and expand transversally much less than a laser pulse of typical parameters diffracts, which allows much longer acceleration distances in PWFA.
- Laser pulses quickly diffract when focused strongly, which allows the generation of locally confined hot spots for tunneling ionization and electron release.
- LWFA allow various injection mechanisms, supported by the low wake velocities due to the laser pulse group velocity, but are for the same reason prone to unwanted self-injection and dark current.
- Injection thresholds for PWFA are comparably high, due to the high wake velocities of dephasing free systems, but, on the other hand, PWFA for the same reason allows realizing dark-current free systems.
- LWFA can be realized in ultracompact, lab-scale setups, but not yet at highest repetition rates and stability.
- Linac-driven PWFA requires large facilities, but can provide bunches with high stability at high repetition rate.
- LWFA inherently generates beams with very high currents, but not extremely low energy spreads, both due to the small plasma cavity sizes and large field gradients.

These features, advantages and disadvantages of LWFA and PWFA are highly complementary.

5.2 Hybrid Combinations of LWFA and PWFA

The complementary features of laser beams, particle beams and LWFA and PWFA can be exploited. For example, using a laser pulse to pre-ionize the plasma for a PWFA stage is an obvious method to harness both the ability of laser pulses to tunnel ionize at comparably low intensities, and to make use of the long, dephasing-free acceleration distances achievable by PWFA. However, there are many other

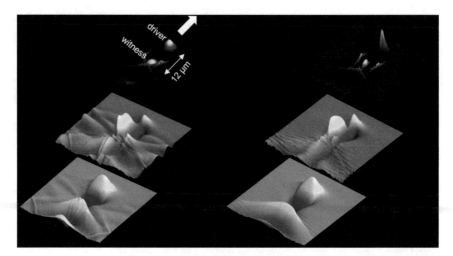

Fig. 5.7 An electron double bunch from an LWFA stage is put into a second, higher density plasma stage, where it acts as a driver-witness pair PWFA energy afterburner. The right hand side shows a later snapshot during the PWFA acceleration, where plasma lensing has pinched both driver and witness bunch. The top plots shows the electron density, the middle plots the transverse electric field, and the bottom plots show the longitudinal wakefields. Visualization adapted from [4]

interesting hybrid permutations, and combinations of hybrid building blocks, which are more complicated than LWFA or PWFA alone, but allow achieving extremely high electron beam quality and tunability. A fundamental subset result of the above discussion and comparison of features of LWFA and PWFA is that:

- High-current electron beams are ideal drivers for plasma waves.
- Laser pulses are ideal to produce dense, high current electron bunches in compact setups.

It is therefore a very attractive option to design and optimize an LWFA stage such that the electron output can be harnessed to drive an attached PWFA stage. This principle has been suggested to exploit purposefully in [4], where "Monoenergetic Energy Doubling in a Hybrid Laser-Plasma Wakefield Accelerator" was proposed. A more or less sharp transition from LWFA to PWFA may occur also in a single plasma stage [5–7] when dephasing is reached and/or when laser pulse power depletion and/or diffraction sets in. Figure 5.7 shows, based on PIC-simulations with OOPIC, how an electron double bunch from an LWFA stage is energy boosted in a driver-witness type PWFA stage.

The key requirement for an electron bunch to drive a strong PWFA wakefield stage is its current. The energy spread of electron beam drivers is less important, because as described above for sufficiently relativistic electrons the phase slippage or bunch lengthening is small and not a primary concern. On the contrary, it is well-known that in a process called BNS damping [8] an energy spread can suppress instabilities such as the beam breakup (BBU) instability and hosing. While the energy transfer of the drive beam to the plasma will introduce an energy spread even if initially perfectly monoenergetic, it helps if the drive beam has a significant energy spread right from the beginning of the PWFA process [9]. This is a feature which LWFA can realize very well: produce inherently ultrashort, multi-kA electron bunches [10, 11, 41] with significant energy spreads in very compact setups. Significant energy spread of LWFA electron output is a drawback or even showstopper for example for demanding applications such as the free-electron laser, but in contrast, for PWFA, the significant energy spread can be even an asset.

Experimentally, plasma beam dumps via collective deceleration [13]—a signature of energy transfer from driver beam to plasma and as such a first step in the direction of hybrid LWFA-PWFA—has been observed in a setup with two gas jets [14]. Passive plasma lensing has also been shown experimentally in a similar setup [15]. This is important, because one of the drawbacks of LWFA-generated electron bunches (in fact, of any plasma wakefield-accelerated electron bunch) is the typically large divergence with which they leave the plasma stage. This means that: (a) the divergence has to be reduced during the extraction process from the LWFA stage as much as possible; and (b) the electron beam needs to be captured by a transport line soon after the first stage. Long down-ramps which allow adiabatic extraction of the electron beam help with regard to Point (a), and with regard to Point (b) passive plasma lensing as in [15] can be exploited during the transition from LWFA to PWFA [16]—a highly attractive option, because at the same time it pinches the electron beam, which in turn increases its $\tilde{Q}$ and hence strengthens the interaction with the plasma. Other options are active plasma lenses, or if possible, capture by strong permanent magnet quadrupoles.

To control LWFA-PWFA staging, and to extend the acceleration distance in the PWFA stage, e.g., in view of head erosion, the PWFA stage needs to be preionized. In [15], the diffracting remnant of the LWFA laser pulse was used—as long as one does not exceed the dephasing limit, this laser pulse fraction arrives in the PWFA stage earlier than the LWFA-generated electron beam and can tunnel ionize the gas in the PWFA stage. The laser remnant intensity has to exceed the tunneling ionization threshold, but should also not be too high, otherwise the plasma electrons are heated. Selective full ionization of hydrogen, but not helium, was observed in [15]. However, the laser pulse diffraction of the remnant laser pulse cannot be mitigated, and the distance over which suitable ionization is achieved is limited. The laser intensity at this point is below the relativistic self-focusing threshold, so that this mitigation mechanism is not present. On the contrary, ionization defocusing may play a significant role, which further reduces the length over which the LWFA remnant pulse may be useful for ionizing the PWFA stage. Long-range preionization, for example with an axilens in counterpropagating geometry as suggested in [17], is required over and beyond an initial ionization distance provided by the LWFA remnant, in order to fully unlock the long acceleration distances which PWFA enables.

Next to acceleration as a goal, hybrid LWFA-PWFA systems do also allow to investigate various basic PWFA-specific features and challenges in compact setups, such as PWFA plasma dynamics and ion motion [18] or may allow innovative light sources applications [1, 19].

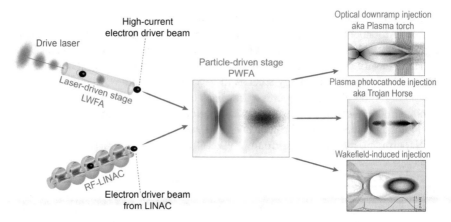

Fig. 5.8 Schematic overview on: linac-driven PWFA (**top left**); hybrid LWFA+PWFA (**bottom left**); and three different PWFA injection schemes (**right**)

In addition to huge energy gains in single accelerator stages, the low electric fields of the PWFA driver allows realizing a unique set of electron injection methods, such as plasma photocathodes also known as Trojan Horse [20, 21], wakefield-induced ionization (WII) injection [22], and optical density downramp injection also known as plasma torch [23]. These methods allow boosting the quality of electron bunches by many orders of magnitude, and therefore may pave the way to high performance key applications, such as for hard X-ray light sources or for potential building blocks for high energy physics research.

Figure 5.8 summarizes various options of injection on the right hand side. The central gateway building block here is PWFA, either driven by electron beams from linacs, or from LWFA.

In the following, the three injection methods are discussed briefly. In addition to the general idea of utilizing LWFA-generated bunches for PWFA, these are currently explored in the Strathclyde/DESY-led Work Package 14: "Hybrid Laser-electron-beam driven acceleration" of EuPRAXIA, the EU H2020 design study for a European Plasma Research Accelerator with eXcellence In Applications (2015–2019). Hybrid LWFA-PWFA and the Trojan Horse method, specifically, has been supported by RadiaBeam Technology's US DOE-funded "Plasma Photocathode Beam Brightness Transformer for Laser-Plasma Wakefield Accelerators" (2013–2016) in a Strathclyde–RadiaBeam–UCLA-centered collaboration. Here, the work was broken down into the plasma photocathode research on the one hand, which was developed within the "E210: Trojan Horse PWFA" experimental programme (2012–2017) at the SLAC FACET linac, and R&D on the exploitation of LWFA-generated electron bunches for PWFA on the other hand. Figure 5.8 visualizes the underlying conceptual approaches. A high-current electron beam, either coming from a linac, or from an LWFA stage, drives a PWFA stage, in which ionization injection based methods may generate electron beams of superior brightness, boosted energy, etc. These approaches are more complex when compared to single-stage LWFA, as well as linac-driven PWFA. However, they offer both compactness and high quality output beams, and are increasingly seen as pathway towards substantially higher beam quality from lab-scale accelerators as an alternative to conventional methods of bunch generation, which have limits that arise from fundamental principles. Schematically, one may draw a parallel to how many modern high power lasers operate: just as in state-of-the-art Ti:sapphire lasers green pump laser pulses are used to generate infrared laser pulses of much higher power and eventually, intensity, electron beams of already high density are used to generate electron bunches of much higher phase space density and brightness with the approaches sketched below.

5.2.1 Plasma Torch—All Optical Density Downramp Injection

Density downramp injection is an attractive method of bunch generation both for LWFA and PWFA, as it can realize localized and tunable injection of electrons into the plasma wave. It relies on a localized elongation of the plasma cavity on the plasma density downrampp, which facilities injection and trapping of plasma electrons into the plasma cavity. Density downramp injection is a state-of-the-art injection method for LWFA, where modern implementations even allow increasingly stable generation even of double bunches [24], which could be used, e.g., for electron energy afterburners [4]. While the use of a sharp density downramp for localized injection had been first suggested for PWFA [25], it has not been realized for PWFA until recently in context of the E210 collaboration at SLAC FACET (to be published). One reason for this is the much poorer availability of PWFA driver beams when compared to high-power laser pulses for LWFA. Another one is the higher Lorentz factor γ of electron beam driven plasma waves compared to LWFA, which increases the threshold of injection as pointed out above, and a further one is the general practical complexity and difficulty when generating downramps hydrodynamically, in particular in multi-component gas mixtures. However, in such multi-component mixtures with gases with lower and higher ionization thresholds, or in PWFA which relies on self-ionization of the supporting gas/plasma medium, plasma downramps can be generated optically, as suggested in [23, 26]. In this approach, the plasma density spike and downramp are generated by an additional laser pulse, which generates plasma in the path of the electron beam driver by tunneling ionization of a gas component with higher ionization threshold. For example, in a hydrogen–helium gas mixture, preionized hydrogen can be used to support the wakefield, and helium with its substantially higher ionization threshold is ionized by a laser pulse which generates a plasma column or "plasma torch" perpendicular to the driver electron beam path on top of the preionized hydrogen. Figure 5.9 visualizes this approach by particle-in-cell simulations with the code VSim. The electron beam driver (black) propagates to the right through a hydrogen plasma channel, and encounters a perpendicular

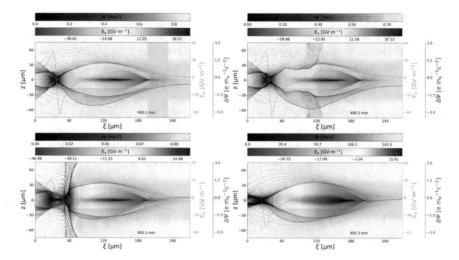

Fig. 5.9 Plasma torch injection in PIC-simulations with VSim. The drive beam (black) propagates to the right and drives a plasma wave based on preionized hydrogen. A laser pulse has generated a helium-based plasma torch perpendicular to the drive beam propagation axis, constitutes a density ramp and triggers density downramp injection

helium plasma torch (grey black). This plasma torch constitutes sharp local plasma density up- and downramps, and, when the electron beam driver arrives, this leads to a local distortion of the plasma wave such that plasma electrons are robustly injected at the desired position in the laboratory frame. Tuning of the laser pulse parameters and helium density allows exploring and optimizing [27] this all-optically-triggered plasma downramp injection in a wide parameter range, including the production of high-brightness electron witness beams.

Owing to the extremely steep density downramp gradients, which this method allows to produce, injection can be achieved with comparably low driver beam currents of the order of 1 kA. Timing of the plasma torch injection method is uncritical, because the plasma torch filament, being generated by a laser pulse just above the ionization threshold, is rather cold and does not change much over a wide timing window of at least tens of ps, ultimately until recombination sets in. It is important to emphasize that the laser pulse required to generate the optical density spike can be of sub-mJ class for typical pulse durations of tens of femtoseconds, as the intensity required to tunnel ionize even helium sufficiently, the element with highest (first) ionization threshold, for Ti:sapphire lasers amounts to few 10^{15} W/cm^2. At these intensities and corresponding power levels, laser pulses are far more manageable than those with relativistic intensities such as required for LWFA. The feasibility of this approach is therefore very high, and as regards laser pulse management substantially less demanding than, e.g., double laser pulse and/or staged LWFA approaches. The required laser pulse powers to generate suitable plasma torches is even borderline within reach of fibre or thin disc laser systems. Shaping of the transverse laser pulse intensity profile, in combination with large tunability of the gas ratio, allows to produce plasma density spikes and downramps in a wide parameter range, including also very high charge beams, and with possibly very high stability. As with conventional downramp injection methods, the injected electron beam population is automatically produced on axis and therefore automatically aligned with the driver beam, which is an advantage of this injection method. As with other downramp injection methods, the injection rate is not only dependent on the shape of the downramp, but also on the strength and shape of the driver pulse and the corresponding wakefield strength and blowout sheath trajectories, which may vary from shot-to-shot in particular when using electron output from LWFA as drivers, which at present shows substantial shot-to-shot variations. This coupling is a disadvantage of this otherwise highly doable and attractive method. A first realization of this method—and at the same time the first realization of any downramp injection scheme in PWFA—has been achieved during the E210 programme at SLAC FACET (to be published). Once the necessary infrastructure was installed, the plasma torch downramp injection could be established within few days of beamtime in a reliable manner, which confirms the expectation of high feasibility of this method as implied by theory.

5.2.2 Trojan Horse—Plasma Photocathode Injection

The Trojan Horse [28], also known as plasma photocathode injection method, is related to plasma torch injection. However, here the laser pulse needs to arrive very shortly after the electron beam driver, such that additional electrons via ionization of the higher ionization threshold medium are released approximately in the center of the blowout. Then, these electrons, which initially have negligible residual momentum due to the low laser intensity, are then rapidly accelerated in the forward direction by the plasma wakefield, which allows realizing particularly low emittance values.

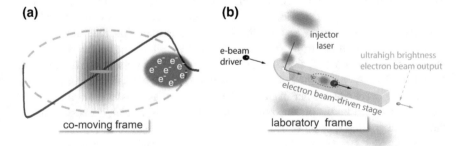

Fig. 5.10 **a** Collinear Trojan Horse schematic in the co-moving frame; and **b** laboratory frame. The laser focusing pulse releases ultracold electrons (green) from the high ionization threshold medium, which then briefly fall back and are trapped in the back part of the accelerating blowout

This can be achieved in various geometries between driver electron beam and release laser [20], e.g., in perpendicular or collinear geometry [21].

Figure 5.10 visualizes for collinear geometry how the process works in co-moving (left) frame and laboratory frame (right). The laser pulse focuses to intensities exceeding the ionization threshold of the higher ionization threshold medium (e.g., helium) in the center of the electron beam driven blowout, and releases helium electrons, represented by the green ellipse, at arbitrary position within the blowout. The center of the blowout is particularly interesting, because here the longitudinal wakefield is zero and the corresponding trapping potential is maximized: electrons released here harness the full accelerating field of the wakefield, which optimizes the use of the driver-excited wakefield for trapping. It is also an ideal position because here, longitudinal and transverse plasma wakefields are zero, and hence the laser-controlled release rate of helium electrons is completely decoupled from the wakefield excitation process. The witness bunch generation is therefore much more decoupled from drive beam jitter than in other schemes. One can also release with transverse longitudinal offset at other positions within the blowout, simply by shifting the release laser pulse with respect to the wakefield center. If one releases farther away from the center, however, transverse and longitudinal plasma wakefields may have to be taken into account for the tunneling ionization rates as superposition of laser and plasma wakefields. Because the trapping potential parabolic shape is flat at its maximum in the middle of the blowout, this position is also robust against longitudinal (i.e., timing) jitter of the release laser pulse. This may be important in context of linac-driven Trojan Horse, where there is no intrinsic synchronization between electron and laser beam. Finally, the center blowout position is also favourable because it allows to realize a robust dark-current free PWFA, when the trapping region is confined to the area around the center of the blowout [29].

The plasma photocathode therefore allows uniquely tunable electron bunch generation. As regards obtainable beam quality, the key advantage is that the release laser pulse requires intensities just above the ionization threshold of comparably low ionization levels. In practice, e.g., a Ti:sapphire laser pulse at an intensity of $\approx 10^{15}$ W/cm^2 is sufficient to liberate electrons from helium, similar as for plasma torch injection. The ponderomotive force of such a laser pulse is comparably low, which in turn means that the residual transverse momentum of the released electrons is low—the released electrons are not transversally pushed out by the laser pulse, but simply remain on axis, fall back and catch up with the plasma wave. This is in contrast to LWFA, where the laser pulse *has* to be intense enough to expel electrons off axis, and to excite the plasma wave. The low residual transverse momentum of electrons released by the plasma photocathode process is crucial, because it means the thermal emittance of produced electrons is very small [21, 30].

The release laser can in principle be realized at arbitrary frequency [20, 28], e.g., frequency-multiplied. This could be useful because then the electric field E_0 which is needed for tunneling ionization is reached at a lower laser intensity, due to the $E_0 = a_0 \times 3.2 \times 10^{-12}$ V/m $/\lambda[\mu\text{m}]$ scaling with the normalized light amplitude a_0. Therefore, if operating at frequency-doubled laser light, i.e., at $\lambda \approx 400$ nm, a_0 can be decreased by a factor of 2, which further reduces the residual transverse momentum and thermal emittance contribution. The use of higher frequency light also has impact on the minimum obtainable spot size of a Gaussian laser pulse, and on the Rayleigh length, which has to be taken into account when calculating thermal emittance and phase mixing. However, at higher frequencies, multi-photon-ionization becomes relevant [31], which may limit the range of applicability of shorter wavelength injector pulses, and of all-optical two-colour variations such as mentioned in [32, 33] and explored in [34]. The latter approaches have a different set of challenges and limitations, e.g., as regards dephasing (in particular for longer wavelength λ driver pulses due to $L_d \approx \lambda_p^3/\lambda^2$) and diffraction of the driver beam, practical limitations in context of drive beam laser availability, tunability and parameter range, and may require gas-dynamic confinement of the high-ionization threshold component as in [34].

Key advantages of the Trojan Horse method are:

- The far-reaching decoupling of wakefield excitation from laser-based ionization injection with low-power laser pulses offers robustness and tunability, including use of different gas species for wakefield excitation and injection, respectively.
- The unprecedented range of emittance and brightness of the obtainable electron beam: A key factor here is the residual transverse momentum of the released electrons, which means the (thermal) emittance scales with laser spot size and the normalized release laser amplitude a_0 [21, 30].

Because electrons can be released in the center of the blowout, a minimum drive beam current of 5–6 kA is sufficient to allow for trapping. Such currents are straightforwardly obtainable as output from LWFA systems. Such an LWFA-PWFA-TH system constitutes a triple-hybrid approach, as it would make use of the laser system for generating the electron beam driver in LWFA, for preionization of the PWFA stage, and for witness bunch generation. Figure 5.11 visualizes this potential setup. The preionization laser here is focused by an axilens and applied in counter-propagating geometry to allow a well-defined and wide plasma channel, and the plasma photocathode release laser has intrinsic synchronization with the electron drive beam—this hybrid setup therefore harnesses fully the advantages of being initially LWFA-driven.

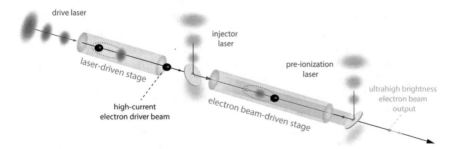

Fig. 5.11 Potential setup of a "triple-hybrid" system where the laser system generates the electron bunch driver for PWFA via LWFA, is used to preionize the PWFA stage, and for witness bunch generation via the Trojan Horse mechanism

Trojan Horse allows for tens of nm-rad scale normalized emittance values, kA-scale currents and hence in combination for unprecedented 5D-brightness. By employing tailored beam loading mechanisms [17], the energy spread of the produced electron bunches can potentially be controlled and reduced to sub-0.01% values in one and the same plasma stage. This is important, as this reduces the challenges associated with extraction and transport of electron beams from the plasma stage and for beam quality and emittance preservation substantially. A low energy spread is also crucial for key applications, e.g., for photon science drivers.

It should be noted that the added level of complexity of hybrid LWFA-PWFA in various configurations when compared to, e.g., single-stage LWFA must not be confused with limited feasibility! On the contrary, the Trojan Horse scheme, which promises highest output beam quality and tunability, is composed such that the individual building blocks are all well controllable due to threshold effects: The preionization of the PWFA stage requires deploying laser pulse intensity sufficient to ionize the low ionization threshold component such as hydrogen, and excess intensity does not change the produced local plasma yield once 100% of the hydrogen is ionized. This acts as a bandpass filter against shot-to-shot laser intensity jitter.

Further, by operating at low plasma densities and hence large blowouts, the release position of electrons inside the blowout can be very stable and tunable: for example, for a blowout of $\approx 500\,\mu$m size, the center of the blowout can be hit reliably with better than 1% stability shot-by-shot when assuming a plasma photocathode laser versus electron driver beam time-of-arrival (and pointing) jitter of <30 fs. A sensitivity analysis of realistic timing, pointing and laser intensity jitter (to be published) confirms the expectation that near-constant charge bunches with tens of nmrad normalized emittance values and corresponding ultrahigh brightness values can stably be achieved shot-by-shot. When operating at such reduced plasma densities, the accelerating fields still amount to tens of GV/m, and next to injection stability the moderate fields are also advantageous because the reduced field gradient inside the blowout leads to a reduced residual energy spread [17]. While ≈ 30 fs laser-electron beam timing stability can be reached at state-of-the-art linac-driven FEL facilities, using a split-off laser pulse from the LWFA system in the hybrid approach can assumedly be delivered with a synchronization stability at or better than the 1-fs level.

In the hybrid LWFA-PWFA variant, at present the major sources of shot-to-shot-jitter of the LWFA output are pointing, energy, energy spread, charge and current. However, energy spread and energy jitter is rather uncritical for any of the hybrid approaches for reasons discussed above, and as regards driver current jitter, the Trojan Horse scheme has unique advantages due to the injection process being to a large degree decoupled from a varying wakefield strength and shape shot-by-shot. Even if the drive beam current jitters substantially from shot-to-shot, and as a direct consequence the excited wakefield strength, the plasma photocathode-produced witness beam charge is completely independent from this, as it is only a function of the laser pulse and high-ionization-threshold medium (e.g., helium) density. Due to the parabolic shape of the trapping potential around the center of the blowout and the corresponding longitudinal electric wakefield, this laser-gated injection process has a further auto-stabilizing function even in case of drive beam shot-to-shot jitter. The approach therefore combines prospects for highest beam quality with ultrahigh tunability as well as potentially very high stability compared to other plasma wakefield approaches. The first demonstration of (linac-driven) Trojan Horse could be realized at SLAC FACET in the E210 programme, using plasma torch injection as a stepping stone (to be published). This was realized under boundary conditions of incoming beam jitter as well as blowout size which have been rather unfavourable and will be much improved e.g., for SLAC FACET-II, which confirms and fosters expectations as regards controllability and impact of this method in the future.

5.2.3 Wakefield-Induced Ionization—WII

The wakefield-induced ionization (WII) injection method [22] exploits PWFAs operating at a high transformer ratio in order to induce ionization and trapping of high-quality electron bunches from and into the extreme accelerating fields of the plasma wake. These electrons originate from wakefield-induced ionization over an atomic species with appropriate ionization threshold, which is doped into the background plasma in a short axial region of the plasma target.

Electrons from the dopant species not only need to be ionized by the accelerating wakefields, but they also need to be trapped by the wakefields before they reach the end of the blowout cavity. This establishes a necessary condition for trapping which can be expressed in terms of the difference in wakefield potential between the initial and final phase positions of the electrons within the plasma wake [35], i.e., $\psi_i - \psi_f = 1$, where ψ is the normalized wakefield potential, related to the longitudinal wakefields by $E_z = (mc^2/e) \, \partial_\zeta \psi$. Thus, the generated wakefields must be strong enough to generate a maximum difference in ψ greater than one. This imposes a constraint on the peak current of the drive beam, which needs to be at least 5 kA. This is a fundamental limit on the current of the drive beam required to generate a wakefield capable of trapping electrons originated from ionization [36].

In particular for WII injection, the drive beams need an even higher peak current (around 10 kA or higher), in order to generate a stronger wakefield capable of trapping within a shorter phase range. However, high-current drive beams could also induce ionization over the dopant species by means of its space charge fields, if they are narrow enough. Therefore, in order to avoid ionization from the fields of the beam, it is necessary to start the WII injection process with a relatively wide drive beam. In this way, the trapped electrons will be originated by WII only, thereby constraining the final phase-space volume occupied by the witness beam, and, consequently, increasing its final quality. Remarkably, starting the plasma-wakefield generation process with a relatively wide drive beam also allows for largely improved stability of the PWFA system [37].

Figure 5.12 shows an example of WII injection performed with the PIC code OSIRIS [38] in 3D Cartesian geometry, and considering a drive beam with similar parameters to those attainable in the FLASHForward experiment [36]. In the simulation, a Gaussian drive beam with a peak current of $I_b^0 = 10$ kA, a longitudinal

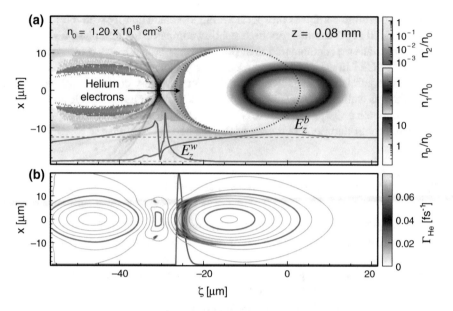

Fig. 5.12 OSIRIS 3D simulation of a PWFA with WII injection: **a** electron density of the drive beam, the plasma and the ionized helium; and **b** ionization rate for He according to the ADK model. The contours in (**b**) show the wakefield-equipotential surfaces in steps of 0.2

(rms) size of $\sigma_z = 7$ μm, a transverse (rms) size of $\sigma_x = 4$ μm is initialized with an average energy of 1 GeV. The total charge of the beam is $Q = 575$ pC. In Fig. 5.12, the drive beam is traversing a homogeneous plasma at the resonant density $n_0 = 1.2 \times 10^{18}$ cm^{-3}. The plasma is doped with neutral helium at 1% concentration. Near the end of the blowout cavity the accelerating wakefields are high enough to ionize completely the first electronic level of helium (Fig. 5.12b). From the total phase-space volume of ionized electrons, only the ones closer to the axis can be trapped within the blowout, provided that a difference in potential $\psi_i - \psi_f = 1$ can be asymptotically reached at the same time that the acquired transverse oscillation amplitude is smaller than the radial extent of the blowout at the final phase position ζ_f. This constrains the initial phase-space volume of the trapped electrons to a thin disc centered on axis which extends up to a radial position r_{max}. An upper estimate of the normalized transverse emittance of the so-injected witness bunch can be given in terms of the initial transverse extent of this disc r_{max} [39], i.e., $k_p \epsilon_n^{max} = (k_p r_{max})^2/8$ [36]. Since typically $k_p r_{max}$ is smaller than one, a practical rule of thumb to scale the final normalized emittance of the WII-injected witness beams as a function of the plasma density can be given as $\epsilon_n^{max} \approx 0.1/k_p = 0.5$ μm$/\sqrt{n_p[10^{18} \text{ cm}^{-3}]}$.

WII injection also allows for beam-loading optimization [40]. The magnitude of the current profile of the injected beams can be controlled by adjusting the concentration of the dopant species in a similar fashion to how it is done in LWFAs with ionization injection [41]. The magnitude of the current profile required for a minimal time correlated energy spread along the witness bunch is higher for higher transformer ratio cases [36]. Therefore, since WII injection requires of a high-transformer ratio by design, the witness beams needs a high current (few tens of kiloamps) for an optimal beam-loading, and, consequently, for a low correlated energy spread [36].

In summary, WII injection allows for the production of high current, low energy spread and low emittance electron beams, which can be accelerated to double or thrice the initial energy of the drive beam, all in a conceptually particularly simple setup.

Utilizing LWFA-generated electron beams to perform a PWFA with WII injection has unique advantages:

- Electron beams from LWFAs can be of high current, reaching peaks of around 30 kA and beyond [41]. Thus, they fulfill the essential requirement to enable WII injection in the PWFA stage. Other typical features of LWFA beams such as a ∼10% relative energy spread or a relatively high normalized emittance of ∼5 μm are even beneficial to use them as drivers, given the enhanced stabilization that they provide to the PWFA system [9, 37].
- WII injection is triggered by the wakefields themselves, and therefore, there is no need for a precise time synchronization of external components, such as lasers, to enable injection. Moreover, the WII injection method is insensitive to the jittering of the pointing angle of the electron beam emerging from the previous LWFA.
- Electron beams from LWFAs are typically of short duration, 10–20 fs [41]. This means that the corresponding resonant density reaches values near 10^{19} cm^{-3}. At these densities the produced witness beams feature normalized emittances values below 100 nm and sub-fs duration.
- It is best for the performance of WII injection to have relatively wide drive beams entering the plasma stage in order to avoid injection from the drive beam fields and to provide improved stability [37]. This means that LWFA to PWFA staging with WII injection can tolerate a certain drift in vacuum of the highly divergent electron beam from LWFA. This simplifies the LWFA to PWFA setup as there is potentially no need of beam optics between stages to transport and refocus the LWFA beam into the PWFA stage.

5.3 Applications

The above mentioned schemes can allow production of electron beams with dramatically improved quality in compact hybrid setups. An improvement in electron beam quality, in particular normalized emittance ϵ_n, and corresponding output electron brightness $B \propto I/\epsilon_n^2$ by orders of magnitude, and substantial decrease of relative energy spread down to the <0.01% level already at few GeV energies may be possible in a single stage. This has fundamental impact, for example on the feasibility of light sources such as free-electron lasers. Driving a free-electron-laser with electron output from plasma-based accelerators is one of the main goals of the plasma accelerator community. However, there are strict requirements on transverse phase space (emittance) and longitudinal phase space (energy and energy spread) of the FEL-driving electron beam in order to achieve lasing, which are so far prohibitive for realizing this goal with conventional LWFA-based approaches:

- The emittance criterion: The emittance of the produced electron beam determines which resonant FEL wavelength λ_r may be realizable at given electron energy γ via $\epsilon_n < \lambda_r \gamma / 4\pi$ (also known as Pellegrini criterion). For a normalized emittance two orders of magnitude better than state-of-the-art, a hard X-ray FEL could be realized already at few GeV electron energies. For example, at $\epsilon_n < 50$ nm, a hard X-ray lasing at $\lambda_r \approx 1.5$ Å can be achieved already at an electron beam energy of 2 GeV.
- The energy spread criterion: The relative energy spread σ_γ / γ of the produced electron beam must be much smaller than the FEL Pierce parameter ρ, which for a hard X-ray FEL means σ_γ / γ has to be better than 0.1%. Typical plasma accelerators produce beams with energy spread in the percent range, but, e.g., the plasma photocathode technique in an augmented version [17] will allow the production of beams with relative energy spread even down to the 0.01% range.
- The FEL gain: The 1D FEL gain length scales with brightness B as $L_{g,1D} \propto B_{5D}^{-1/3}$ and because $B \propto \epsilon_n^{-2}$, e.g., a two orders of magnitude lower emittance and hence four orders of magnitude higher brightness means a much higher FEL gain, and shorter gain length, can be realized with such beams. This allows shrinking down the FEL undulator section to the ten-meter scale [21] instead of hundreds of meters as today, and may even allow for unprecedentedly ultrashort single spike, high brilliance coherent hard X-ray pulses.

The features of the hybrid schemes described above therefore may allow transformative impact on future compact, high performance FELs on multiple levels via dramatically improved emittance, energy spread and brightness, and hence may allow hard X-ray FEL's to become ubiquitous, and to achieve higher performance. Other light sources such as inverse Compton scattering or betatron radiation/ion channel lasers would likewise profit from such enhanced electron beam quality. Figure 5.13

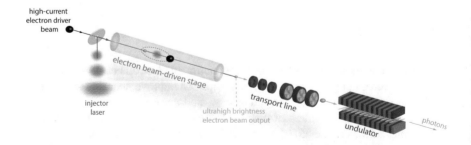

Fig. 5.13 Schematic of an undulator-based FEL driven by the output of a hybrid plasma accelerator based on LWFA electron beams, with a plasma photocathode PWFA stage as beam brightness converter

visualizes the principle setup of an FEL driven by a hybrid plasma wakefield accelerator (here, by a collinear Trojan Horse system).

5.4 Summary

The fundamental motivation for realizing advanced hybrid plasma accelerators is discussed. LWFA electron beam output (large intrinsic multi-kA-scale currents and significant energy spreads) is a very attractive candidate to drive PWFA stages, which in turn allow realizing advanced witness bunch generation techniques. These techniques may allow the production of tunable electron bunches with unprecedented quality, which paves the way to key applications, e.g., for photon science. An increasing number of groups is therefore investigating both hybrid LWFA-PWFA plasma accelerators, e.g., with high current beams from LWFA such as those shown in [41] as well as advanced injection schemes. The conceptual elegance of these novel approaches, and the many recent successes as regards first experimental demonstrations, which have been obtained with comparably limited available resources and beamtimes, may indicate that hybrid LWFA-PWFA is a highly attractive path to substantially increased beam quality such as brightness as well as regards tunability and stability. It may therefore develop into a key contribution to fulfill and unleash the decade-old promise of plasma accelerators as future transformative particle and radiation sources.

Funding This research was funded by H2020 EuPRAXIA (Grant No. 653782), ERC (Grant No. 715807), EPSRC (Grant No. EP/N028694/1), DFG (MAP EXC 158), DFG Emmy-Noether (B.H.), and used computational resources of the National Energy Research Scientific Computing Center, which is supported by DOE DE-AC02-05CH11231, and of Shaheen (Project k1191).

Conflicts of Interest The authors declare no conflict of interest.

References

1. B. Hidding, G.G. Manahan, O. Karger, A. Knetsch, G. Wittig, D.A. Jaroszynski, Z.M. Sheng, Y. Xi, A. Deng, J.B. Rosenzweig et al., Ultrahigh brightness bunches from hybrid plasma accelerators as drivers of 5th generation light sources. J. Phys. B **47**, 234010 (2014)
2. C. Nieter, J.R. Cary, VORPAL: a versatile plasma simulation code. J. Comput. Phys. **196**, 448–473 (2004). https://doi.org/10.1016/j.jcp.2003.11.004
3. N. Barov, J.B. Rosenzweig, M.C. Thompson, R.B. Yoder, Energy loss of a high-charge bunched electron beam in plasma: analysis. Phys. Rev. ST Accel. Beams **7**, 061301 (2004). https://doi.org/10.1103/PhysRevSTAB.7.061301
4. B. Hidding, T. Koenigstein, J. Osterholz, S. Karsch, O. Willi, G. Pretzler, Monoenergetic energy doubling in a hybrid laser-plasma wakefield accelerator. Phys. Rev. Lett. **104**, 195002 (2010). https://doi.org/10.1103/PhysRevLett.104.195002

5. F.S. Tsung, R. Narang, W.B. Mori, C. Joshi, R.A. Fonseca, L.O. Silva, Near-GeV-energy laser-wakefield acceleration of self-injected electrons in a centimeter-scale plasma channel. Phys. Rev. Lett. **93**, 185002 (2004). https://doi.org/10.1103/PhysRevLett.93.185002
6. K.H. Pae, I.W. Choi, J. Lee, Self-mode-transition from laser wakefield accelerator to plasma wakefield accelerator of laser-driven plasma-based electron acceleration. Phys. Plasmas **17**, 123104 (2010). https://doi.org/10.1063/1.3522757
7. P.E. Masson-Laborde, M.Z. Mo, A. Ali, S. Fourmaux, P. Lassonde, J.C. Kieffer, W. Rozmus, D. Teychenné, R. Fedosejevs, Giga-electronvolt electrons due to a transition from laser wakefield acceleration to plasma wakefield acceleration. Phys. Plasmas **21**, 123113 (2014). https://doi.org/10.1063/1.4903851
8. V.E. Balakin, A.V. Novokhatsky, V.P. Smirnov, Vlepp: transverse beam dynamics. Conf. Proc. **C830811**, 119–120 (1983)
9. T.J. Mehrling, R.A. Fonseca, A. Martinez de la Ossa, J. Vieira, Mitigation of the hose instability in plasma-wakefield accelerators. Phys. Rev. Lett. **118**, 174801 (2017). https://doi.org/10.1103/PhysRevLett.118.174801
10. S.M. Wiggins, R.C. Issac, G.H. Welsh, E. Brunetti, R.P. Shanks, M.P. Anania, S. Cipiccia, G.G. Manahan, C. Aniculaesei, B. Ersfeld et al., High quality electron beams from a laser wakefield accelerator. Plasma Phys. Control. Fusion **52**, 124032–124039 (2010). https://doi.org/10.1088/0741-3335/52/12/124032
11. O. Lundh, J. Lim, C. Rechatin, L. Ammoura, A. Ben-Ismail, X. Davoine, G. Gallot, J.P. Goddet, E. Lefebvre, V. Malka et al., Few femtosecond, few kiloampere electron bunch produced by a laser-plasma accelerator. Nat. Phys. **7**, 219–222 (2011)
12. J. Couperus, R. Pausch, A. Köhler, O. Zarini, J. Krämer, M. Garten, A. Huebl, R. Gebhardt, U. Helbig, S. Bock et al., Demonstration of a beam loaded nanocoulomb-class laser wakefield accelerator. Nat. Commun. **8**, 487 (2017)
13. H.C. Wu, T. Tajima, D. Habs, A.W. Chao, J. Meyer-ter Vehn, Collective deceleration: toward a compact beam dump. Phys. Rev. ST Accel. Beams **13**, 101303 (2010). https://doi.org/10.1103/PhysRevSTAB.13.101303
14. S. Chou, J. Xu, K. Khrennikov, D.E. Cardenas, J. Wenz, M. Heigoldt, L. Hofmann, L. Veisz, S. Karsch, Collective deceleration of laser-driven electron bunches. Phys. Rev. Lett. **117**, 144801 (2016)
15. S. Kuschel, D. Hollatz, T. Heinemann, O. Karger, M.B. Schwab, D. Ullmann, A. Knetsch, A. Seidel, C. Rödel, M. Yeung et al., Demonstration of passive plasma lensing of a laser wakefield accelerated electron bunch. Phys. Rev. Accel. Beams **19**, 071301 (2016). https://doi.org/10.1103/PhysRevAccelBeams.19.071301
16. T. Heinemann, A. Knetsch, O. Zarini, A. Martinez de la Ossa, T. Kurz, O. Kononenko, U. Schramm, A. Köhler, B. Hidding, A. Irman, et al., Investigating the key parameters of a staged laser- and particle driven plasma wakefield accelerator experiment, in *Proceedings of the 8th International Particle Accelerator Conference (IPAC 2017)* (Copenhagen, Denmark, 2017), p. TUPIK010. https://doi.org/10.18429/JACOW-IPAC2017-TUPIK010
17. G. Manahan, A. Habib, P. Scherkl, P. Delinikolas, A. Beaton, A. Knetsch, O. Karger, G. Wittig, T. Heinemann, Z. Sheng et al., Single-stage plasma-based correlated energy spread compensation for ultrahigh 6D brightness electron beams. Nat. Commun. **8**, 15705 (2017)
18. M.F. Gilljohann, H. Ding, A. Döpp, J. Götzfried, S. Schindler, G. Schilling, S. Corde, A. Debus, T. Heinemann, B. Hidding et al., Direct observation of plasma waves and dynamics induced by laser-accelerated electron beams. Phys. Rev. X **9**, 011046 (2019). https://doi.org/10.1103/PhysRevX.9.011046
19. J. Ferri, S. Corde, A. Döpp, A. Lifschitz, A. Doche, C. Thaury, K. Ta Phuoc, B. Mahieu, I.A. Andriyash, V. Malka, et al., High-brilliance betatron γ-ray source powered by laser-accelerated electrons. Phys. Rev. Lett. **120**, 254802 (2018). https://doi.org/10.1103/PhysRevLett.120.254802
20. B. Hidding, G. Pretzler, D. Bruhwiler, J. Rosenzweig, Method for generating electron beams in a hybrid plasma accelerator. German Patent DE 10 2011 104 858.1 (2011)

21. B. Hidding, G. Pretzler, J.B. Rosenzweig, T. Königstein, D. Schiller, D.L. Bruhwiler, Ultracold electron bunch generation via plasma photocathode emission and acceleration in a beam-driven plasma blowout. Phys. Rev. Lett. **108**, 035001 (2012). https://doi.org/10.1103/PhysRevLett.108.035001
22. A. Martinez de la Ossa, J. Grebenyuk, T. Mehrling, L. Schaper, J. Osterhoff, High-quality electron beams from beam-driven plasma accelerators by wakefield-induced ionization injection. Phys. Rev. Lett. **111**, 245003 (2013). https://doi.org/10.1103/PhysRevLett.111.245003
23. G. Wittig, O. Karger, A. Knetsch, Y. Xi, A. Deng, J.B. Rosenzweig, D.L. Bruhwiler, J. Smith, G.G. Manahan, Z.M. Sheng et al., Optical plasma torch electron bunch generation in plasma wakefield accelerators. Phys. Rev. ST Accel. Beams **18**, 081304 (2015). https://doi.org/10.1103/PhysRevSTAB.18.081304
24. J. Wenz, K. Khrennikov, A. Döpp, M. Gilljohann, H. Ding, J. Goetzfried, S. Schindler, A. Buck, J. Xu, M. Heigoldt, et al., Tunable femtosecond electron and X-ray double-beams from a compact laser-driven accelerator (2018). arXiv:1804.05931
25. H. Suk, N. Barov, J.B. Rosenzweig, E. Esarey, Plasma electron trapping and acceleration in a plasma wake field using a density transition. Phys. Rev. Lett. **86**, 1011–1014 (2001). https://doi.org/10.1103/PhysRevLett.86.1011
26. G. Wittig, O.S. Karger, A. Knetsch, Y. Xi, A. Deng, J.B. Rosenzweig, D.L. Bruhwiler, J. Smith, Z.M. Sheng, D.A. Jaroszynski, et al., Electron beam manipulation, injection and acceleration in plasma wakefield accelerators by optically generated plasma density spikes. Nucl. Instrum. Methods Phys. Res. Sect. **829**, 83–87 (2016). 2nd European Advanced Accelerator Concepts Workshop-EAAC 2015. https://doi.org/10.1016/j.nima.2016.02.027
27. A. Martinez de la Ossa, Z. Hu, M.J.V. Streeter, T.J. Mehrling, O. Kononenko, B. Sheeran, J. Osterhoff, Optimizing density down-ramp injection for beam-driven plasma wakefield accelerators. Phys. Rev. Accel. Beams **20**, 091301 (2017). https://doi.org/10.1103/PhysRevAccelBeams.20.091301
28. B. Hidding, J.B. Rosenzweig, Y. Xi, B. O'Shea, G. Andonian, D. Schiller, S. Barber, O. Williams, G. Pretzler, T. Königstein et al., Beyond injection: trojan horse underdense photocathode plasma wakefield acceleration. AIP Conf. Proc. **1507**, 570–575 (2012). https://doi.org/10.1063/1.4773760
29. G.G. Manahan, A. Deng, O. Karger, Y. Xi, A. Knetsch, M. Litos, G. Wittig, T. Heinemann, J. Smith, Z.M. Sheng et al., Hot spots and dark current in advanced plasma wakefield accelerators. Phys. Rev. Accel. Beams **19**, 011303 (2016). https://doi.org/10.1103/PhysRevAccelBeams.19.011303
30. C.B. Schroeder, J.L. Vay, E. Esarey, S.S. Bulanov, C. Benedetti, L.L. Yu, M. Chen, C.G.R. Geddes, W.P. Leemans, Thermal emittance from ionization-induced trapping in plasma accelerators. Phys. Rev. ST Accel. Beams **17**, 101301 (2014). https://doi.org/10.1103/PhysRevSTAB.17.101301
31. Y. Xi, B. Hidding, D. Bruhwiler, G. Pretzler, J.B. Rosenzweig, Hybrid modeling of relativistic underdense plasma photocathode injectors. Phys. Rev. ST Accel. Beams **16**, 031303 (2013). https://doi.org/10.1103/PhysRevSTAB.16.031303
32. D. Umstadter, J.K. Kim, E. Dodd, Method and apparatus for generating and accelerating ultrashort electron pulses. U.S. Patent 5,789,876 (1995)
33. D. Umstadter, J.K. Kim, E. Dodd, Laser injection of ultrashort electron pulses into wakefield plasma waves. Phys. Rev. Lett. **76**, 2073–2076 (1996). https://doi.org/10.1103/PhysRevLett.76.2073
34. L.L. Yu, E. Esarey, C. Schroeder, J.L. Vay, C. Benedetti, C. Geddes, M. Chen, W. Leemans, Two-color laser-ionization injection. Phys. Rev. Lett. **112**, 125001 (2014). https://doi.org/10.1103/PhysRevLett.112.125001
35. E. Oz, S. Deng, T. Katsouleas, P. Muggli, C.D. Barnes, I. Blumenfeld, F.J. Decker, P. Emma, M.J. Hogan, R. Ischebeck et al., Ionization-induced electron trapping in ultrarelativistic plasma wakes. Phys. Rev. Lett. **98**, 084801 (2007). https://doi.org/10.1103/PhysRevLett.98.084801
36. A. Martinez de la Ossa, T.J. Mehrling, L. Schaper, M.J.V. Streeter, J. Osterhoff, Wakefield-induced ionization injection in beam-driven plasma accelerators. Phys. Plasmas **22**, 093107 (2015)

37. A.M. de la Ossa, T. Mehrling, J. Osterhoff, Intrinsic stabilization of the drive beam in plasma wakefield accelerators. Phys. Rev. Lett. **121**, 064803 (2018)
38. R. Fonseca, L. Silva, F. Tsung, V. Decyk, W. Lu, C. Ren, W. Mori, S. Deng, S. Lee, T. Katsouleas et al., OSIRIS: a three-dimensional, fully relativistic particle in cell code for modeling plasma based accelerators. Lecture Notes in Computer Science (Springer, Berlin, 2002), pp. 342–351
39. N. Kirby, I. Blumenfeld, C.E. Clayton, F.J. Decker, M.J. Hogan, C. Huang, R. Ischebeck, R.H. Iverson, C. Joshi, T. Katsouleas et al., Transverse emittance and current of multi-GeV trapped electrons in a plasma wakefield accelerator. Phys. Rev. ST Accel. Beams **12**, 051302 (2009)
40. M. Tzoufras, W. Lu, F.S. Tsung, C. Huang, W.B. Mori, T. Katsouleas, J. Vieira, R.A. Fonseca, L.O. Silva, Beam loading in the nonlinear regime of plasma-based acceleration. Phys. Rev. Lett. **101**, 145002 (2008). https://doi.org/10.1103/PhysRevLett.101.145002
41. J. Couperus, R. Pausch, A. Köhler, O. Zarini, J. Krämer, M. Garten, A. Huebl, R. Gebhardt, U. Helbig, S. Bock et al., Demonstration of a beam loaded nanocoulomb-class laser wakefield accelerator. Nat. Commun. **8**, 487 (2017). https://doi.org/10.1038/s41467-017-00592-7

Chapter 6
Introduction to High Brightness Electron Beam Dynamics

M. Ferrario

Abstract In this paper we introduce, from basic principles, the main concepts of beam focusing and transport of space charge dominated beams in high brightness accelerators using the beam envelope equation as a convenient mathematical tool. Matching conditions suitable for preserving beam quality are derived from the model for significant beam dynamics regimes.

6.1 Introduction

Light sources based on high-gain free electron lasers or future high-energy linear colliders require the production, acceleration, and transport up to the interaction point of low divergence, high charge-density electron bunches [1]. Many effects contribute in general to degradation of the final beam quality, including chromatic effects, wake fields, emission of coherent radiation, and accelerator misalignments. Space charge effects and mismatch with focusing and accelerating devices typically contribute to emittance degradation of high charge-density beams [2]; hence, control of beam transport and acceleration is the leading edge for high-quality beam production.

Space charge effects represent a very critical issue and a fundamental challenge for high-quality beam production and its applications. Without proper matching, significant emittance growth may occur when the beam is propagating through different stages and components owing to the large differences of transverse focusing strength. This unwanted effect is even more serious in the presence of finite energy spread.

In this paper we introduce, from basic principles, the main concepts of beam focusing and transport in modern accelerators using the beam envelope equation as a convenient mathematical tool. Matching conditions suitable for preserving beam quality are derived from the model for significant beam dynamics regimes. A more detailed discussion of the previous topics can be found in the many classical textbooks on this subject, as listed in [3–6].

M. Ferrario (✉)
Frascati National Laboratory, National Institute for Nuclear Physics, Rome, Italy
e-mail: Massimo.Ferrario@lnf.infn.it

© Springer Nature Switzerland AG 2019

L. A. Gizzi et al. (eds.), *Laser-Driven Sources of High Energy Particles and Radiation*,
Springer Proceedings in Physics 231, https://doi.org/10.1007/978-3-030-25850-4_6

6.2 Laminar and Non-laminar Beams

An ideal high-charge particle beam has orbits that flow in layers that never intersect, as occurs in a laminar fluid. Such a beam is often called a laminar beam. More precisely, a laminar beam satisfies the following two conditions [6]:

(i) all particles at a given position have identical transverse velocities. On the contrary, the orbits of two particles that start at the same position could separate and later cross each other;
(ii) assuming that the beam propagates along the z axis, the magnitudes of the slopes of the trajectories in the transverse directions x and y, given by $x'(z) = \mathrm{d}x/\mathrm{d}z$ and $y'(z) = \mathrm{d}y/\mathrm{d}z$, are linearly proportional to the displacement from the z axis of beam propagation.

Trajectories of interest in beam physics are always confined to the inside of small, near-axis regions, and the transverse momentum is much smaller than the longitudinal momentum, $p_{x,y} \ll p_z \approx p$. As a consequence, it is possible in most cases to use the small angle, or *paraxial*, approximation, which allows us to write the useful approximate expressions $x' = p_x/p_z \approx p_x/p$ and $y' = p_y/p_z \approx p_x/p$.

To help understand the features and advantages of a laminar beam propagation, the following figures compare the typical behaviour of a laminar and a non-laminar (or thermal) beam.

Figure 6.1 illustrates an example of orbit evolution of a laminar mono-energetic beam with half width x_0 along a simple beam line with an ideal focusing element

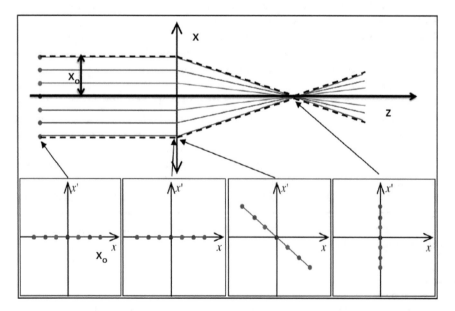

Fig. 6.1 Particle trajectories and phase space evolution of a laminar beam [19]

(solenoid, magnetic quadrupoles, or electrostatic transverse fields are usually adopted to this end), represented by a thin lens located at the longitudinal coordinate $z = 0$. In an ideal lens, focusing (defocusing) forces are linearly proportional to the displacement from the symmetry axis z, so that the lens maintains the laminar flow of the beam.

The beam shown in Fig. 6.1 starts propagating completely parallel to the symmetry axis z; in this particular case, the particles all have zero transverse velocity. There are no orbits that cross each other in such a beam. Ignoring collisions and inner forces, such as Coulomb forces, a parallel beam could propagate an infinite distance with no change in its transverse width. When the beam crosses the ideal lens, it is transformed into a converging laminar beam. Because the transverse velocities after the linear lens are proportional to the displacement off axis, particle orbits define similar triangles that converge to a single point. After passing through the singularity at the focal point, the particles follow diverging orbits. We can always transform a diverging (or converging) beam into a parallel beam by using a lens of the proper focal length, as can be seen by reversing the propagation axis of Fig. 6.1.

The small boxes in the lower part of the figure depict the particle distributions in the trace space (x, x'), equivalent to the canonical phase space $(x, p_x \approx x'p)$ when p is constant, i.e., without beam acceleration. The phase space area occupied by an ideal laminar beam is a straight segment of zero thickness. As can be easily verified, the condition that the particle distribution has zero thickness proceeds from condition (6.1); the segment straightness is a consequence of condition (6.2). The distribution of a laminar beam propagating through a transport system with ideal linear focusing elements is thus a straight segment with variable slope.

Particles in a non-laminar beam have a random distribution of transverse velocities at the same location and a spread in directions, as shown in Fig. 6.2. Because of the disorder of a non-laminar beam, it is impossible to focus all particles from a location in the beam toward a common point. Lenses can influence only the average motion of particles. Focal spot limitations are a major concern for a wide variety of applications, from electron microscopy to free electron lasers and linear colliders. The phase space plot of a non-laminar beam is no longer a straight line: the beam, as shown in the lower boxes of Fig. 6.2, occupies a wider area of the phase space.

6.3 The Emittance Concept

The phase space surface A occupied by a beam is a convenient figure of merit for designating the quality of a beam. This quantity is the emittance ε_x and is usually represented by an ellipse that contains the whole particle distribution in the phase space (x, x'), such that $A = \pi\varepsilon_x$. An analogous definition holds for the (y, y') and (z, z') planes. The original choice of an elliptical shape comes from the fact that when linear focusing forces are applied to a beam, the trajectory of each particle in phase space lies on an ellipse, which may be called the trajectory ellipse. Being the area of

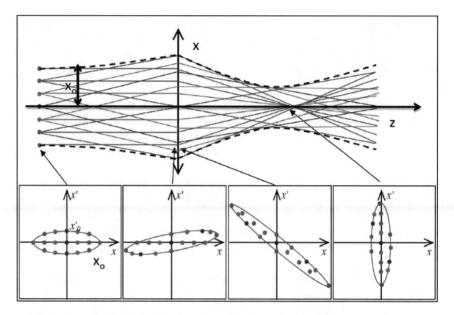

Fig. 6.2 Particle trajectories and phase space evolution of a non-laminar beam [19]

the phase space, the emittance is measured in metres radians. More often is expressed in millimetres milliradians or, equivalently, in micrometres.

The ellipse equation is written as

$$\gamma_x x^2 + 2\alpha_x x x' + \beta_x x'^2 = \varepsilon_x, \tag{6.1}$$

where x and x' are the particle coordinates in the phase space and the coefficients $\alpha_x(z)$, $\beta_x(z)$, and $\gamma_x(z)$ are called Twiss parameters, which are related by the geometrical condition:

$$\beta_x \gamma_x - \alpha_x^2 = 1 \tag{6.2}$$

As shown in Fig. 6.3, the beam envelope boundary X_{max}, its derivative $(X_{max})'$, and the maximum beam divergence X'_{max}, i.e., the projection on the axes x and x' of the ellipse edges, can be expressed as a function of the ellipse parameters:

$$\begin{cases} X_{max} = \sqrt{\beta_x \varepsilon_x} \\ (X_{max})' = -\alpha \sqrt{\frac{\varepsilon}{\beta}} \\ X'_{max} = \sqrt{\gamma_x \varepsilon_x} \end{cases} \tag{6.3}$$

According to Liouville's theorem, the six-dimensional (x, p_x, y, p_y, z, p_z) phase space volume occupied by a beam is constant, provided that there are no dissipative

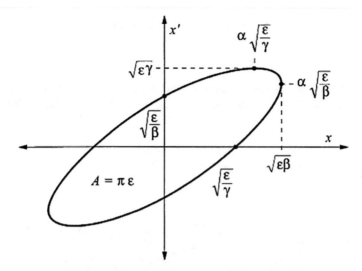

Fig. 6.3 Phase space distribution in a skewed elliptical boundary, showing the relationship of Twiss parameters to the ellipse geometry [6]

forces, no particles lost or created, and no Coulomb scattering among particles. Moreover, if the forces in the three orthogonal directions are uncoupled, Liouville's theorem also holds for each reduced phase space surface, (x, p_x), (y, p_y), (z, p_z), and hence emittance also remains constant in each plane [3].

Although the net phase space surface occupied by a beam is constant, non-linear field components can stretch and distort the particle distribution in the phase space, and the beam will lose its laminar behaviour. A realistic phase space distribution is often very different from a regular ellipse, as shown in Fig. 6.4.

We introduce, therefore, a definition of emittance that measures the beam quality rather than the phase space area. It is often more convenient to associate a statistical definition of emittance with a generic distribution function $f(x, x', z)$ in the phase

Fig. 6.4 Typical evolution of phase space distribution (black dots) under the effects of non-linear forces with the equivalent ellipse superimposed (red line)

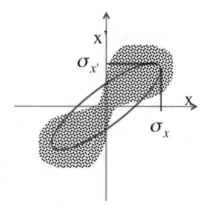

space; this is the so-called *root mean square (rms) emittance*:

$$\gamma_x x^2 + 2\alpha_x xx' + \beta_x x'^2 = \varepsilon_{x,\text{rms}}. \tag{6.4}$$

The rms emittance is defined such that the equivalent-ellipse projections on the x and x' axes are equal to the rms values of the distribution, implying the following conditions:

$$\begin{cases} \sigma_x = \sqrt{\beta_x \varepsilon_{x,\text{rms}}} \\ \sigma_{x'} = \sqrt{\gamma_x \varepsilon_{x,\text{rms}}} \end{cases}, \tag{6.5}$$

where

$$\begin{cases} \sigma_x^2(z) = \langle x^2 \rangle = \int\limits_{-\infty}^{+\infty}\int\limits_{-\infty}^{+\infty} x^2 f(x, x', z)\, dx\, dx' \\ \sigma_{x'}^2(z) = \langle x'^2 \rangle = \int\limits_{-\infty}^{+\infty}\int\limits_{-\infty}^{+\infty} x'^2 f(x, x', z)\, dx\, dx' \end{cases} \tag{6.6}$$

are the second moments of the distribution function $f(x, x', z)$. Another important quantity that accounts for the degree of (x, x') correlations is defined as

$$\sigma_{xx'}(z) = \langle xx' \rangle = \int\limits_{-\infty}^{+\infty}\int\limits_{-\infty}^{+\infty} xx' f(x, x', z)\, dx\, dx'. \tag{6.7}$$

From (6.3) it also holds that

$$\sigma_x' = \frac{\sigma_{xx'}}{\sigma_x} = -\alpha_x \sqrt{\frac{\varepsilon_{x,\text{rms}}}{\beta_x}}$$

See also (6.16), which allows us to link the correlation moment, (6.7), to the Twiss parameter as

$$\sigma_{xx'} = -\alpha_x \varepsilon_{x,\text{rms}}. \tag{6.8}$$

One can easily see from (6.3) and (6.5) that

$$\alpha_x = -\frac{1}{2}\frac{d\beta_x}{dz}$$

also holds.

By substituting the Twiss parameter defined by (6.5) and (6.8) into condition (6.2) we obtain [5]

$$\frac{\sigma_{x'}^2}{\varepsilon_{x,\text{rms}}}\frac{\sigma_x^2}{\varepsilon_{x,\text{rms}}} - \left(\frac{\sigma_{xx'}}{\varepsilon_{x,\text{rms}}}\right) = 1. \tag{6.9}$$

Reordering the terms in (6.8) we obtain the definition of *rms emittance* in terms of the second moments of the distribution:

$$\varepsilon_{\text{rms}} = \sqrt{\sigma_x^2 \sigma_{x'}^2 - \sigma_{xx'}^2} = \sqrt{(\langle x^2 \rangle \langle x'^2 \rangle - \langle xx' \rangle^2)}, \tag{6.10}$$

where we omit, from now on, the subscript x in the emittance notation: $\varepsilon_{\text{rms}} = \varepsilon_{x,\text{rms}}$. The rms emittance tells us some important information about phase space distributions under the effect of linear or non-linear forces acting on the beam. Consider, for example, an idealized particle distribution in phase space that lies on some line that passes through the origin, as illustrated in Fig. 6.5.

Assuming a generic correlation of the type $x' = Cx^n$ and computing the rms emittance according to (6.10) we have

$$\varepsilon_{\text{rms}}^2 = C\sqrt{\langle x^2 \rangle \langle x^{2n} \rangle - \langle x^{n+1} \rangle^2} \begin{cases} n = 1 \Rightarrow \varepsilon_{\text{rms}} = 0 \\ n > 1 \Rightarrow \varepsilon_{\text{rms}} \neq 0 \end{cases}. \tag{6.11}$$

When $n = 1$, the line is straight and the rms emittance is $\varepsilon_{\text{rms}} = 0$. When $n > 1$ the relationship is non-linear, the line in phase space is curved, and the rms emittance is, in general, not zero. Both distributions have zero area. Therefore, we conclude that even when the phase space area is zero, if the distribution is lying on a curved line, its rms emittance is not zero. The rms emittance depends not only on the area occupied by the beam in phase space, but also on distortions produced by non-linear forces.

If the beam is subject to acceleration, it is more convenient to use the rms normalized emittance, for which the transverse momentum $p_x = p_z x' = m_0 c \beta \gamma x'$ is used instead of the divergence:

$$\varepsilon_{n,\text{rms}} = \frac{1}{m_0 c} \sqrt{\sigma_x^2 \sigma_{p_x}^2 - \sigma_{xp_x}^2} = \frac{1}{m_0 c} \sqrt{(\langle x^2 \rangle \langle p_x^2 \rangle - \langle xp_x \rangle^2)} = \sqrt{(\langle x^2 \rangle \langle (\beta \gamma x')^2 \rangle - \langle x\beta \gamma x' \rangle^2)} \tag{6.12}$$

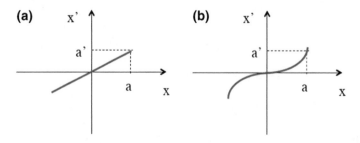

Fig. 6.5 Phase space distributions under the effect of (**a**) linear or (**b**) non-linear forces acting on the beam

The reason for introducing a normalized emittance is that the divergences of the particles $x' = p_x/p$ are reduced during acceleration as p increases. Thus, acceleration reduces the un-normalized emittance, but does not affect the normalized emittance.

It is interesting to estimate the fundamental limit of the beam emittance that is set by quantum mechanics on the knowledge of the two conjugate variables (x, p_x). The state of a particle is actually not exactly represented by a point, but by a small uncertainty volume of the order of $\hbar^3$ in the 6D phase space. According to the Heisenberg uncertainty relation $\sigma_x \sigma_{p_x} \geq \frac{\hbar}{2}$ one gets from (6.12) $\varepsilon_{n,rms}^{QM} \geq \frac{1}{2} \frac{\hbar}{m_o c} = \frac{\lambda_c}{2}$, where λ_c is the reduced Compton wavelength. For electrons it gives:

$$\varepsilon_{n,\text{rms}}^{QM} \geq 1.9 \times 10^{-13} \text{m}.$$

In the classical limit we see also from (6.12) that the single particle emittance is zero.

Assuming a small energy spread within the beam, the normalized and un-normalized emittances can be related by the approximated relation $\langle \beta \gamma \rangle \varepsilon_{\text{rms}}$. This approximation, which is often used in conventional accelerators, may be strongly misleading when adopted for describing beams with significant energy spread, like those currently produced by plasma accelerators. A more careful analysis is reported next [7].

When the correlations between the energy and transverse positions are negligible (as in a drift without collective effects), (6.12) can be written as

$$\varepsilon_{n,rms}^2 = \langle \beta^2 \gamma^2 \rangle \langle x^2 \rangle \langle x'^2 \rangle - \langle \beta \gamma \rangle^2 \langle xx' \rangle^2 \tag{6.13}$$

Consider now the definition of relative energy spread

$$\sigma_\gamma^2 = \frac{\langle \beta^2 \gamma^2 \rangle - \langle \beta \gamma \rangle^2}{\langle \beta \gamma \rangle^2}$$

which can be inserted into (6.13) to give

$$\varepsilon_{n,\text{rms}}^2 = \langle \beta^2 \gamma^2 \rangle \sigma_\gamma^2 \langle x^2 \rangle \langle x'^2 \rangle + \langle \beta \gamma \rangle^2 \left(\langle x^2 \rangle \langle x'^2 \rangle - \langle xx' \rangle^2 \right). \tag{6.14}$$

Assuming relativistic particles ($\beta = 1$), we get

$$\varepsilon_{n,\text{rms}}^2 = \langle \gamma^2 \rangle \left(\sigma_\gamma^2 \sigma_x^2 \sigma_{x'}^2 + \varepsilon_{rms}^2 \right). \tag{6.15}$$

If the first term in the parentheses is negligible, we find the conventional approximation of the normalized emittance as $\langle \gamma \rangle \varepsilon_{rms}$. For a conventional accelerator, this might generally be the case. Considering, for example, beam parameters for the SPARC_LAB photoinjector [8]: at 5 MeV the ratio between the first and the second term is ~10^{-3}; while at 150 MeV it is ~10^{-5}. Conversely, using typical beam parameters at the plasma–vacuum interface, the first term is of the same order of magnitude as for conventional accelerators at low energies; however, owing to the rapid increase of the bunch size outside the plasma ($\sigma_{x'} \sim$ mrad) and the large energy spread ($\sigma_\gamma > 1\%$), it becomes predominant compared with the second term after a drift of a few millimetres. *Therefore, the use of approximated formulas when measuring the normalized emittance of plasma accelerated particle beams is inappropriate* [9].

6.4 The Root Mean Square Envelope Equation

We are now interested in following the evolution of the particle distribution during beam transport and acceleration. One can use the collective variable defined in (6.6), the second moment of the distribution termed the rms beam envelope, to derive a differential equation suitable for describing the rms beam envelope dynamics [10]. To this end, let us compute the first and second derivative of σ_x [4]:

$$\frac{d\sigma_x}{dz} = \frac{d}{dz}\sqrt{\langle x^2 \rangle} = \frac{1}{2\sigma_x}\frac{d}{dz}\langle x^2 \rangle = \frac{1}{2\sigma_x}2\langle xx' \rangle = \frac{\sigma_{xx'}}{\sigma_x}$$

$$\frac{d^2\sigma_x}{dz^2} = \frac{d}{dz}\frac{\sigma_{xx'}}{\sigma_x} = \frac{1}{\sigma_x}\frac{d\sigma_{xx'}}{dz} - \frac{\sigma_{xx'}^2}{\sigma_x^3} = \frac{1}{\sigma_x}\left(\langle x'^2 \rangle + \langle xx' \rangle\right) - \frac{\sigma_{xx'}^2}{\sigma_x^3} = \frac{\sigma_{x'}^2 + \langle xx'' \rangle}{\sigma_x} - \frac{\sigma_{xx'}^2}{\sigma_x^3}.$$

$$(6.16)$$

Rearranging the second derivative in (6.16), we obtain a second-order non-linear differential equation for the beam envelope evolution,

$$\sigma_x'' = \frac{\sigma_x^2\sigma_{x'}^2 - \sigma_{xx'}^2}{\sigma_x^3} + \frac{\langle xx'' \rangle}{\sigma_x},\tag{6.17}$$

or, in a more convenient form, using the rms emittance definition (6.10),

$$\sigma_x'' - \frac{1}{\sigma_x}\langle xx'' \rangle = \frac{\varepsilon_{rms}^2}{\sigma_x^3}.\tag{6.18}$$

In (6.18), the emittance term can be interpreted physically as an outward pressure on the beam envelope produced by the rms spread in trajectory angle, which is parameterized by the rms emittance.

Let us now consider, for example, the simple case with $\langle xx'' \rangle = 0$, describing a beam drifting in free space. The envelope equation reduces to

$$\sigma_x^3\sigma_x'' = \varepsilon_{rms}^2.\tag{6.19}$$

With initial conditions σ_0, σ_0' at z_0, depending on the upstream transport channel, (6.19) has a hyperbolic solution:

$$\sigma(z) = \sqrt{\left(\sigma_0 + \sigma_0'(z - z_0)\right)^2 + \frac{\varepsilon_{rms}^2}{\sigma_0^2}(z - z_0)^2}. \qquad (6.20)$$

Considering the case of a beam at waist $\left(xx' = 0\right)$ with $\sigma_0' = 0$, using (6.5), the solution (6.20) is often written in terms of the β function as

$$\sigma(z) = \sigma_0 \sqrt{1 + \left(\frac{z - z_0}{\beta_w}\right)^2}. \qquad (6.21)$$

This relation indicates that without any external focusing element the beam envelope increases from the beam waist by a factor $\sqrt{2}$ with a characteristic length $\beta_w = \sigma_0^2 / \varepsilon_{rms}$, as shown in Fig. 6.6.

At the waist, the relation $\varepsilon_{rms}^2 = \sigma_{0,x}^2 \sigma_{0,x'}^2$ also holds, which can be inserted into (6.20) to give $\sigma_x^2(z) = \sigma_{0x'}^2 (z - z_0)^2$. Under this condition, (6.15) can be written as

$$\varepsilon_{n,rms}^2(z) = \langle \gamma^2 \rangle \left(\sigma_\gamma^2 \sigma_{x'}^4 (z - z_0)^2 + \varepsilon_{rms}^2\right),$$

showing that beams with large energy spread and divergence undergo a significant normalized emittance growth even in a drift of length $(z - z_0)$ [7, 11].

Notice also that the solution (6.21) is exactly analogous to that of a Gaussian light beam for which the beam width $w = 2\sigma_{ph}$ increases away from its minimum value at the waist w_0 with characteristic length $Z_R = \pi w_0^2 / \lambda$ (Rayleigh length) [4]. This analogy suggests that we can identify an effective emittance of a photon beam as $\varepsilon_{ph} = \lambda / 4\pi$.

For the effective transport of a beam with finite emittance, it is mandatory to make use of some external force providing beam confinement in the transport or accelerating line. The term $\langle xx' \rangle$ accounts for external forces when we know x'', given by the single particle equation of motion:

$$\frac{dp_x}{dt} = F_x. \qquad (6.22)$$

Fig. 6.6 Schematic representation of the beam envelope behaviour near the beam waist

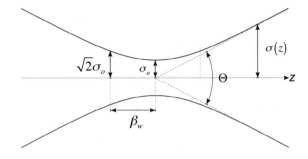

Under the paraxial approximation $p_x \ll p = \beta \gamma mc$, the transverse momentum p_x can be written as $p_x = px' = \beta \gamma m_0 cx'$, so that

$$\frac{dp_x}{dt} = \frac{d}{dt}(px') = \beta c \frac{d}{dz}(px') = F_x, \tag{6.23}$$

and the transverse acceleration results in

$$x'' = -\frac{p'}{p}x' + \frac{F_x}{\beta cp}. \tag{6.24}$$

It follows that

$$\langle xx'' \rangle = -\frac{p'}{p}\langle xx' \rangle + \frac{\langle x F_x \rangle}{\beta cp} = \frac{p'}{p}\sigma_{xx'} + \frac{\langle x F_x \rangle}{\beta cp}. \tag{6.25}$$

Inserting (6.25) into (6.18) and recalling (6.16), $\sigma'_x = \sigma_{xx'}/\sigma_x$, the complete rms envelope equation is:

$$\sigma''_x + \frac{p'}{p}\sigma'_x - \frac{1}{\sigma_x}\frac{\langle x F_x \rangle}{\beta cp} = \frac{\varepsilon^2_{n,rms}}{\gamma^2 \sigma^3_x}, \tag{6.26}$$

where we have included the normalized emittance $\varepsilon_{n,rms} = \gamma \varepsilon_{rms}$. Notice that the effect of longitudinal accelerations appears in the rms envelope equation as an oscillation damping term, called 'adiabatic damping', proportional to p'/p. The term $\langle x F_x \rangle$ represents the moment of any external transverse force acting on the beam, such as that produced by a focusing magnetic channel.

6.5 External Forces

Let's now consider the case of an external linear force acting on the beam in the form $F_x = \mp kx$. It can be focusing or defocusing, according to the sign. The moment of the force is

$$\langle x F_x \rangle = \mp k \langle x^2 \rangle = \mp k\sigma^2_x \tag{6.27}$$

and the envelope equation becomes

$$\sigma''_x + \frac{\gamma'}{\gamma}\sigma'_x \mp k^2_{ext}\sigma_x = \frac{\varepsilon^2_{n,rms}}{\gamma^2 \sigma^3_x}, \tag{6.28}$$

where we have explicitly used the momentum definition $p = \gamma mc$ for a relativistic particle with $\beta \approx 1$ and defined the wavenumber

$$k_{ext}^2 = \frac{k}{\gamma m_0 c^2}.$$

Typical focusing elements are quadrupoles and solenoids [3]. The magnetic quadrupole field is given in Cartesian coordinates by

$$\begin{cases} B_x = B_0 \frac{y}{d} = B_0' y \\ B_y = B_0 \frac{x}{d} = B_0' x \end{cases}, \tag{6.29}$$

where d is the pole distance and B_0' is the field gradient. The force acting on the beam is $\vec{F}_\perp = q v_z B_0' \left(y \hat{j} - x \hat{i} \right)$ and, when B_0 is positive, is focusing in the x direction and defocusing in the y direction. The focusing strength is

$$k_{quad} = \frac{q B_0'}{\gamma m_0 c} = k_{ext}^2.$$

In a solenoid the focusing strength is given by

$$k_{sol} = \left(\frac{q B_0}{2 \gamma m_0 c} \right)^2 = k_{ext}^2.$$

Notice that the solenoid is always focusing in both directions, an important property when the cylindrical symmetry of the beam must be preserved. However, being a second-order quantity in γ, it is more effective at low energy.

It is interesting to consider the case of a uniform focusing channel without acceleration described by the rms envelope equation

$$\sigma_x'' + k_{ext}^2 \sigma_x = \frac{\varepsilon_{rms}^2}{\sigma_x^3}. \tag{6.30}$$

By substituting $\sigma_x = \sqrt{\beta_x \varepsilon_{rms}}$ into (6.30) one obtains an equation for the 'betatron function' $\beta_x(z)$ that is independent of the emittance term:

$$\beta_x'' + 2 k_{ext}^2 \beta_x = \frac{2}{\beta_x} + \frac{\beta_x'^2}{2\beta_x}. \tag{6.31}$$

Equation (6.31) contains just the transport channel focusing strength and, being independent of the beam parameters, suggests that the meaning of the betatron function is to account for the transport line characteristic. The betatron function reflects exterior forces from focusing magnets and is highly dependent on the particular arrangement of the quadrupole magnets. The equilibrium, or matched, solution of (6.31) is given by $\beta_{eq} = \frac{1}{k_{ext}} = \frac{\lambda_\beta}{2\pi}$, as can be easily verified. This result shows that the matched β_x function is simply the inverse of the focusing wavenumber or, equivalently, is proportional to the 'betatron wavelength' λ_β. The corresponding envelope equilibrium condition, i.e., a stationary solution of (6.30), is given by: $\sigma_{\text{eq},x} = \sqrt{\frac{\varepsilon_{\text{rms}}}{k_{\text{ext}}}}$.

In analogy with the kinetic theory of gases we can define the beam temperature in a transverse direction at equilibrium and without correlations as

$$k_B T_{\text{beam},x} = \gamma m_0 \langle v_x^2 \rangle = \frac{\sigma_{p_x}^2}{\gamma m_0} = m_0 c^2 \frac{\varepsilon_{n,\text{rms}}^2}{\gamma \sigma_{\text{eq},x}^2} = \gamma m_0 \beta^2 c^2 \frac{\varepsilon_{\text{rms}}}{\beta_{\text{eq},x}},$$

where k_B is the Boltzmann constant and we have used (6.12), showing that the conditions for a cold beam are typically: low emittance, low energy, high betatron function.

By means of the beam temperature concept one can also define the beam emittance at the source called the thermal emittance. Assuming that electrons are in equilibrium with the cathode temperature $T_c = T_{\text{beam}}$ and $\gamma = 1$, the thermal emittance is given by $\varepsilon_{\text{th,rms}}^{\text{cat}} = \sigma_x \sqrt{\frac{k_B T_c}{m_0 c^2}}$ which, per unit rms spot size at the cathode, is $\varepsilon_{\text{th,rms}} = 0.3\,\mu\text{m/mm}$ at $T_c = 2500$ K. For comparison, in a photocathode illuminated by a laser pulse with photon energy $\hbar\omega$ the expression for the variance of the transverse momentum of the emitted electrons is given by $\sigma_{p_x} = \sqrt{\frac{m_0}{3}(\hbar\omega - \phi_{\text{eff}})}$, where $\phi_{\text{eff}} = \phi_w - \phi_{\text{Schottky}}$, ϕ_w being the material work function and ϕ_{Schottky} the Schottky work function [12]. The corresponding thermal emittance is $\varepsilon_{\text{th,rms}}^{\text{ph}} = \sigma_x \sqrt{\frac{\hbar\omega - \phi_{\text{eff}}}{3 m_0 c^2}}$ that, with the typical parameters of a Copper photocathode illuminated by a UV laser, gives a thermal emittance per unit spot size of about 0.5 μm/mm.

6.6 Space Charge Forces

Another important force acting on the beam is the one produced by the beam itself due to the internal Coulomb forces. The net effect of the Coulomb interaction in a multiparticle system can be classified into two regimes [3]:

(i) *collisional regime*, dominated by binary collisions caused by close particle encounters;
(ii) *collective regime* or *space charge regime*, dominated by the self-field produced by the particles' distribution, which varies appreciably only over large distances compared with the average separation of the particles.

A measure for the relative importance of collisional versus collective effects in a beam with particle density n is the relativistic *Debye length*,

$$\lambda_D = \sqrt{\frac{\varepsilon_0 \gamma^2 k_B T_b}{e^2 n}}. \tag{6.32}$$

As long as the Debye length remains small compared with the particle bunch transverse size, the beam is in the space charge dominated regime and is not sensitive to binary collisions. Smooth functions for the charge and field distributions can be used in this case, and the space charge force can be treated as an external applied force. The space charge field can be separated into linear and non-linear terms as a function of displacement from the beam axis. The linear space charge term defocuses the beam and leads to an increase in beam size. The non-linear space charge terms also increase the rms emittance by distorting the phase space distribution. Under the paraxial approximation of particle motion, we can consider the linear component alone. We shall see next that the linear component of the space charge field can also induce emittance growth when correlations along the bunch are taken into account.

For a bunched beam of uniform charge distribution in a cylinder of radius R and length L, carrying a current $\hat{I}$ and moving with longitudinal velocity $v_z = \beta c$, the linear component of the longitudinal and transverse space charge field are given approximately by [13]

$$E_z(\zeta) = \frac{\hat{I}L}{2\pi \varepsilon_0 R^2 \beta c} h(\zeta), \tag{6.33}$$

$$E_r(r, \zeta) = \frac{\hat{I}r}{2\pi \varepsilon_0 R^2 \beta c} g(\zeta), \tag{6.34}$$

The field form factor is described by the functions:

$$h(\zeta) = \sqrt{A^2 + (1 - \zeta)^2} - \sqrt{A^2 + \zeta^2} - |1 - \zeta| + |\zeta| \tag{6.35}$$

$$g(\zeta) = \frac{(1 - \zeta)}{2\sqrt{A^2 + (1 - \zeta)^2}} + \frac{\zeta}{2\sqrt{A^2 + \zeta^2}}, \tag{6.36}$$

where $\zeta = z/L$ is the normalized longitudinal coordinate along the bunch, $\zeta = 0$ being the bunch tail, and $A = R/\gamma L$ is the beam aspect ratio. The field form factors account for the variation of the fields along the bunch and outside the bunch for $\zeta < 0$ and $\zeta > L$. As γ increases, $g(\zeta) \rightarrow 1$ and $h(\zeta) \rightarrow 0$, thus showing that space charge fields mainly affect transverse beam dynamics. It shows also that an energy increase corresponds to a bunch lengthening in the moving frame $L' = \gamma L$, leading to a vanishing longitudinal field component, as in the case of a continuous beam in the laboratory frame.

To evaluate the force acting on the beam, one must also account for the azimuthal magnetic field associated with the beam current, which, in cylindrical symmetry, is given by

$$B_\vartheta = \frac{\beta}{c} E_r.$$

Thus, the Lorentz force acting on each single particle is given by

$$F_r = e(E_r - \beta c B_\vartheta) = e(1 - \beta^2)E_r = \frac{eE_r}{\gamma^2}. \tag{6.37}$$

The attractive magnetic force, which becomes significant at high velocities, tends to compensate for the repulsive electric force. Therefore, space charge defocusing is primarily a non-relativistic effect and decreases as γ^{-2}.

To include space charge forces in the envelope equation, let us start by writing the space charge forces produced by the previous fields in Cartesian coordinates:

$$F_x = \frac{e\hat{I}x}{8\pi\gamma^2\varepsilon_0\sigma_x^2\beta c}g(\zeta) \tag{6.38}$$

Then, computing the moment of the force, we need

$$x'' = \frac{F_x}{\beta cp} = \frac{eIx}{8\pi\varepsilon_0\gamma^3 m_0\beta^3 c^3\sigma_x^2} = \frac{k_{sc}(\zeta)}{(\beta\gamma)^3\sigma_x^2} \tag{6.39}$$

where we have introduced the generalized beam perveance,

$$k_{sc}(\zeta) = \frac{\hat{I}}{2I_A}g(\zeta) \tag{6.40}$$

where $I_A = 4\pi\varepsilon_0 m_0 c^3/e = 17$ kA is the Alfvén current for electrons. Notice that in this case the perveance in (6.40) explicitly depends on the slice coordinate ζ. We can now calculate the term that enters the envelope equation for a relativistic beam,

$$\langle xx'' \rangle = \frac{k_{sc}}{\gamma^3\sigma_x^2}\langle x^2 \rangle = \frac{k_{sc}}{\gamma^3}, \tag{6.41}$$

leading to the complete envelope equation

$$\sigma_x'' + \frac{\gamma'}{\gamma}\sigma_x' + k_{ext}^2\sigma_x = \frac{\varepsilon_{n,rms}^2}{\gamma^2\sigma_x^3} + \frac{k_{sc}}{\gamma^3\sigma_x}. \tag{6.42}$$

From the envelope (6.42), we can identify two regimes of beam propagation: *space charge dominated* and *emittance dominated*. A beam is space charge dominated as long as the space charge collective forces are largely dominant over the emittance pressure. In this regime, the linear component of the space charge force produces a quasi-laminar propagation of the beam, as one can see by integrating one time (6.39) under the paraxial ray approximation $x' \ll 1$.

A measure of the relative importance of space charge effects versus emittance pressure is given by the *laminarity parameter*, defined as the ratio between the space charge term and the emittance term:

$$\rho = \frac{\hat{I}}{2I_A\gamma}\frac{\sigma^2}{\varepsilon_n^2}. \tag{6.43}$$

When ρ greatly exceeds unity, the beam behaves as a laminar flow (all beam particles move on trajectories that do not cross), and transport and acceleration require a careful tuning of focusing and accelerating elements to keep laminarity. Correlated emittance growth is typical in this regime, which can be made reversible if proper beam matching conditions are fulfilled, as discussed next. When $\rho < 1$, the beam is emittance dominated (thermal regime) and space charge effects can be neglected. The transition to the thermal regime occurs when $\rho \approx 1$, corresponding to the transition energy

$$\gamma_{tr} = \frac{\hat{I}}{2I_A}\frac{\sigma^2}{\varepsilon_n^2}. \tag{6.44}$$

For example, a beam with $\hat{I} = 100$ A, $\varepsilon_n = 1$ μm, and $\sigma = 300$ μm is leaving the space charge dominated regime and is entering the thermal regime at the transition energy of 131 MeV. From this example, one may conclude that the space charge dominated regime is typical of low-energy beams. Actually, for such applications as linac-driven free electron lasers, peak currents exceeding kA are required. Space charge effects may recur if bunch compressors are active at higher energies and a new energy threshold with higher $\hat{I}$ must be considered.

6.7 Correlated Emittance Oscillations

When longitudinal correlations within the bunch are important, like that induced by space charge effects, beam envelope evolution is generally dependent also on the coordinate along the bunch ζ. In this case, the bunch should be considered as an ensemble of n longitudinal slices of envelope $\sigma_s(z, \zeta)$, whose evolution can be computed from n slice envelope equations equivalent to (6.42), provided that the bunch parameters refer to each single slice: γ_s, γ_s', $k_{sc,s} = k_{sc}g(\zeta)$. Correlations within the bunch may cause emittance oscillations that can be evaluated, once an analytical or numerical solution [13] of the slice envelope equation is known, by using the following correlated emittance definition:

$$\varepsilon_{rms,cor} = \sqrt{\langle\sigma_s^2\rangle\langle\sigma_s'^2\rangle - \langle\sigma_s\sigma_s'\rangle^2}, \tag{6.45}$$

where the average is performed over the entire slice ensemble, assuming uniform charge distribution within each slice. In the simplest case of a two-slice model, the previous definition reduces to

$$\varepsilon_{rms,cor} = |\sigma_1\sigma_2' - \sigma_2\sigma_1'|, \tag{6.46}$$

which represents a simple and useful formula for an estimation of the emittance scaling [14].

The total normalized rms emittance is given by the superposition of the correlated and uncorrelated terms as

$$\varepsilon_{\text{rms,cor}} = \langle \gamma \rangle \sqrt{\varepsilon_{\text{rms}}^2 + \varepsilon_{\text{rms,cor}}^2}. \tag{6.47}$$

An interesting example to consider here, showing the consequences of non-perfect beam matching, is the propagation of a beam in the space charge dominated regime nearly matched to an external focusing channel, as illustrated in Fig. 6.7. To simplify our computations, we can neglect acceleration, as in the case of a simple beam transport line made by a long solenoid ($k_{\text{ext}}^2 = k_{\text{sol}}$). The envelope equation for each slice, indicated as σ_s, reduces to

$$\sigma_s'' + k_{\text{ext}}^2 \sigma_s = \frac{k_{\text{sc},s}}{\gamma^3 \sigma_s}. \tag{6.48}$$

A stationary solution corresponding to slice propagation with constant envelope, called *Brillouin flow*, is given by

$$\sigma_{r,\text{B}} = \frac{1}{k_{\text{ext}}} \sqrt{\frac{\hat{I} g(\zeta)}{2\gamma^3 I_A}}, \tag{6.49}$$

where the local dependence of the current $\hat{I}_s = \hat{I} g(\zeta)$ within the bunch has been explicitly indicated. This solution represents the matching conditions for which the external focusing completely balances the internal space charge force. Unfortunately, since k_{ext} has a slice-independent constant value, the Brillouin matching condition is different for each slice and usually cannot be achieved at the same time for all of the bunch slices. Assuming that there is a reference slice perfectly matched (6.49) with an envelope $\sigma_{r,\text{B}}$ and negligible beam energy spread, the matching condition for the other slices can be written as:

Fig. 6.7 Schematic representation of a nearly matched beam in a long solenoid. The dashed line represents the reference slice envelope matched to the Brillouin flow condition. The other slice envelopes are oscillating around the equilibrium solution

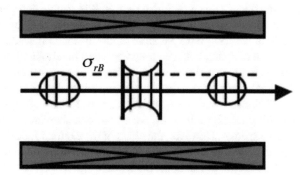

$$\sigma_{s,B} = \sigma_{r,B} + \frac{\sigma_{r,B}}{2}\left(\frac{\delta I_s}{\hat{I}}\right), \tag{6.50}$$

with respect to the reference slice. Considering a slice with a small perturbation δ_s with respect to its own equilibrium (6.50) in the form

$$\sigma_s = \sigma_{s,B} + \delta_s, \tag{6.51}$$

and substituting into (6.48), we can obtain a linearized equation for the slice offset

$$\delta_s'' + 2k_{\text{ext}}^2\delta_s = 0, \tag{6.52}$$

which has a solution given by

$$\delta_s = \delta_0 \cos\left(\sqrt{2}k_{\text{ext}}z\right), \tag{6.53}$$

where $\delta_0 = \sigma_{so} - \sigma_{sB}$ is the amplitude of the initial slice mismatch, which we assume, for convenience, is the same for all slices. Inserting (6.53) into (6.51) we get the perturbed solution:

$$\sigma_s = \sigma_{s,B} + \delta_0 \cos\left(\sqrt{2}k_{\text{ext}}z\right). \tag{6.54}$$

Equation (6.54) shows that slice envelopes oscillate together around the equilibrium solution with the same frequency for all slices ($\sqrt{2}k_{\text{ext}}$, often called the plasma frequency) dependent only on the external focusing forces. This solution represents a collective behaviour of the bunch, similar to that of the electrons subject to the restoring force of ions in a plasma. Using the two-slice model and (6.54), the emittance evolution (6.46) results in

$$\varepsilon_{\text{rms,cor}} = \frac{1}{4}k_{\text{ext}}\sigma_{r,B}\left|\frac{\Delta I}{\hat{I}}\delta_0 \sin(k_{\text{ext}}z)\right|, \tag{6.55}$$

where $\Delta I = \hat{I}_1 - \hat{I}_2$. Notice that, in this simple case, envelope oscillations of the mismatched slices induce correlated emittance oscillations that periodically return to zero, showing the reversible nature of the correlated emittance growth. It is, in fact, the coupling between transverse and longitudinal motion induced by the space charge fields that allows reversibility. With proper tuning of the transport line length or of the focusing field, one can compensate for the transverse emittance growth.

At first, it may seem surprising that a beam with a single charge species can exhibit plasma oscillations, which are characteristic of plasmas composed of two-charge species. However, the effect of the external focusing force can play the role of the other charge species, providing the necessary restoring force that is the cause of such collective oscillations, as shown in Fig. 6.8. The beam can actually be considered as a single-component, relativistic, cold plasma.

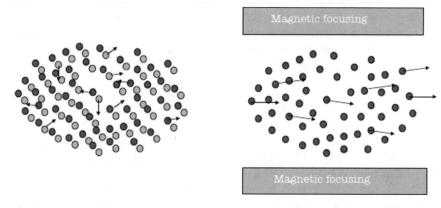

Fig. 6.8 The restoring force produced by the ions (green dots) in a plasma may cause electron (red dots) oscillations around the equilibrium distribution. In a similar way, the restoring force produced by a magnetic field may cause beam envelope oscillations around the matched envelope equilibrium

It is important to bear in mind that beams in linacs are also different from plasmas in some important respects [5]. One is that beam transit time through a linac is too short for the beam to reach thermal equilibrium. Also, unlike a plasma, the Debye length of the beam may be larger than, or comparable to, the beam radius, so shielding effects may be incomplete.

6.8   Matching Conditions in a Radiofrequency Linac

In order to prevent space charge induced emittance growth in a radiofrequency (rf) linac, as in the case of a high brightness photoinjector, and to drive a smooth transition from the space charge to the thermal regime, space charge induced emittance oscillations have to be damped along the linac in such a way that an emittance minimum is obtained at the transition energy (6.44). To this end the beam has to be properly matched to the accelerating sections with a Brillouin like flow in order to keep under control emittance oscillations that in this case are provided by the ponderomotive rf focusing force [2] acting in the rf structures. In some case rf focusing is too weak to provide sufficient beam containment. A long solenoid around the accelerating structure is a convenient replacement to provide the necessary focusing.

The matching conditions for a beam subject to acceleration (assuming $\gamma(z = \gamma_0 + \gamma'z)$ and $\gamma'' = 0$) can be obtained following the previous example (Brillouin flow). This process can be described using the envelope (6.42) for a generic slice σ_s with external focusing provided by $k_{\text{ext}}^2 = k_{\text{sol}} + k_{\text{rf}}^2$, where $k_{\text{rf}}^2 = \frac{\eta}{8}\left(\frac{\gamma'}{\gamma}\right)^2$ and $\gamma' = \frac{eE_{acc}}{mc^2}$ The quantity η is a measure of the higher spatial harmonic amplitudes of the rf wave and it is generally quite close to unity in standing wave (SW) structures and close to 0 in travelling wave (TW) structures [15].

Being now $\gamma(z)$ a time-dependent function, a stationary solution of (6.42) cannot be found by simply looking for a constant envelope solution. A possible way to find an 'equilibrium' solution is described hereafter. By substituting the reduced variable $\hat{\sigma} = \sqrt{\gamma}\sigma_s$ [16] in the envelope (6.42) we obtain

$$\hat{\sigma}'' + \hat{k}_{\text{ext}}^2 \hat{\sigma} = \frac{\hat{K}_{\text{sc}}}{\hat{\sigma}} + \frac{\varepsilon_n^2}{\hat{\sigma}^3} \tag{6.56}$$

with the scaled parameters $\hat{k}_{\text{ext}}^2 = k_{\text{ext}}^2 + \frac{1}{4}\left(\frac{\gamma'}{\gamma}\right)^2 = k_{\text{sol}} + \frac{1}{4}\left(\frac{\gamma'}{\gamma}\right)^2\left(1 + \frac{\eta}{2}\right)$ and $\hat{K}_{\text{sc}} = k_{\text{sc}}/\gamma^2$. Equation (6.56) is equivalent to (6.42) but the damping term has disappeared and the $\hat{k}_{\text{ext}}^2$ and $\hat{K}_{\text{sc}}$ parameters have the same γ^{-2} dependence. In the space charge regime the emittance term can be neglected in (6.56) and an equilibrium solutions in the reduced variables (called the 'invariant envelope' in the literature [2]) is given by $\hat{\sigma}_{\text{sc}} = \frac{\sqrt{\hat{K}_{\text{sc}}}}{\hat{k}_{\text{ext}}}$, corresponding to the matching conditions for the beam envelope:

$$\sigma_{\text{sc}} = \sqrt{\frac{2\hat{I}}{\gamma I_A\left(\Theta^2 + \gamma'^2\left(\frac{\eta}{2} + 1\right)\right)}} \quad \text{for } \rho > 1 \tag{6.57}$$

where $\Theta = \frac{eB}{mc}$.

The expression for the emittance oscillation in the space charge dominated regime, i.e. when $\gamma < \gamma_{\text{tr}}$, can be obtained from (6.55) using reduced variables and results:

$$\varepsilon_n = \frac{1}{\gamma}\sqrt{\frac{\hat{I}}{34 I_A}}\left|\frac{\Delta I}{\hat{I}}\delta_0 \sin\left(\frac{\left(\Theta^2 + \gamma'^2\left(\frac{\eta}{2} + 1\right)\right)^{\frac{1}{2}}}{2\gamma}z\right)\right|. \tag{6.58}$$

Before the transition energy is achieved the emittance performs damped oscillations with wavelength depending on the external fields and with amplitude depending on the current profile. A careful tuning of the external fields and bunch charge profile can minimize the value of the emittance at the injector extraction. A successful application of the emittance compensation technique can be seen in [17, 18].

When the beam enters in the thermal regime an equilibrium solution can be found directly from (6.42) neglecting the space charge term. The result is

$$\sigma_{\text{th}} = \sqrt{\frac{2\varepsilon_n}{\left(\Theta^2 + \frac{\eta}{2}\gamma'^2\right)^{1/2}}} \quad \text{for } \rho < 1 \tag{6.59}$$

and no correlated emittance oscillations are expected. Note also that (6.57) scales like $\gamma^{-1/2}$ while (6.59) is independent of γ.

Acknowledgements I wish to thank A. Cianchi, E. Chiadroni, J.B. Rosenzweig, A.R. Rossi, A. Bacci, and L. Serafini for the many helpful discussions and suggestions.

References

1. T. Shintake, Review of the worldwide SASE FEL development, in *Proceedings of 22nd Particle Accelerator Conference*, Albuquerque, NM (IEEE, New York, 2007), p. 89. https://doi.org/10.1109/PAC.2007.4440331
2. L. Serafini, J.B. Rosenzweig, Phys. Rev. E **55**, 7565 (1997). https://doi.org/10.1103/PhysRevE.55.7565
3. M. Reiser, *Theory and Design of Charged Particle Beams* (Wiley, New York, 1994). https://doi.org/10.1002/9783527617623
4. J.B. Rosenzweig, *Fundamentals of Beam Physics* (Oxford University Press, Oxford, 2003). https://doi.org/10.1093/acprof:oso/9780198525547.001.0001
5. T. Wangler, *Principles of RF Linear Accelerators* (Wiley, New York, 1998). https://doi.org/10.1002/9783527618408
6. S. Humphries, *Charged Particle Beams* (Wiley, New York, 2002)
7. M. Migliorati et al., Phys. Rev. ST Accel. Beams **16**, 011302 (2013). https://doi.org/10.1103/PhysRevSTAB.16.011302
8. M. Ferrario et al., Nucl. Instrum. Methods Phys. Res. B **309**, 183 (2013). https://doi.org/10.1016/j.nimb.2013.03.049
9. A. Cianchi et al., Nucl. Instrum. Methods Phys. Res. A **720**, 153 (2013). https://doi.org/10.1016/j.nima.2012.12.012
10. F.J. Sacherer, IEEE Trans. Nucl. Sci. **NS-18** (1971) 1105. https://doi.org/10.1109/TNS.1971.4326293
11. K. Floettmann, Phys. Rev. ST Accel. Beams **6**, 034202 (2003). https://doi.org/10.1103/PhysRevSTAB.6.034202
12. D.H. Dowell et al., Phys. Rev. ST Accel. Beams **12**, 074201 (2009). https://doi.org/10.1103/PhysRevSTAB.12.074201
13. M. Ferrario et al., Int. J. Mod. Phys. A **22**, 4214 (2007). https://doi.org/10.1142/S0217751X07037779
14. J. Buon, Beam phase space and emittance, in *Proceedings of CERN Accelerator School: 5th General Accelerator Physics Course* (Jyvaskyla, Finland, 1992), CERN-94-01. https://doi.org/10.5170/CERN-1994-001
15. J.B. Rosenzweig, L. Serafini, Phys. Rev. E **49**, 1599 (1994). https://doi.org/10.1103/PhysRevE.49.1599
16. C. Wang, Phys. Rev. E **74**, 046502 (2006). https://doi.org/10.1103/PhysRevE.74.046502
17. R. Akre et al., Phys. Rev. ST Accel. Beams **11**, 030703 (2008). https://doi.org/10.1103/PhysRevSTAB.11.030703
18. M. Ferrario et al., Phys. Rev. Lett. **99**, 234801 (2007). https://doi.org/10.1103/PhysRevLett.99.234801
19. N. Pichoff, Beam dynamics basics in RF linacs, in *Proceedings of CERN Accelerator School: Small Accelerators* (Zeegse, The Netherlands, 2005), CERN-2006-012. https://doi.org/10.5170/CERN-2006-012

Chapter 7
Ion Acceleration: TNSA and Beyond

Marco Borghesi

Abstract This paper reviews experimental progress in laser-driven ion acceleration as well as discussing some of the current and foreseen applications employing laser-accelerated beams of ions. While sheath acceleration processes initiated by high-intensity irradiation of solid foils (the so-called target Normal Sheath Acceleration, *TNSA*) have now been studied for two decades, novel processes which can accelerate ions from the bulk of the irradiated target have emerged more recently. We will summarize the basic physics behind all these mechanisms, as well as briefly reporting current experimental evidence.

7.1 Introduction

The first experiments reporting laser acceleration of protons with beam-like properties and multi-MeV energies were reported in 2000 [1–3]. Experiments since then have demonstrated, over a wide range of laser and target parameters, the generation of multi-MeV proton and ion beams with unique properties such as ultrashort burst emission, high brilliance, and low emittance, which have in turn stimulated ideas for a range of innovative applications. While most of this work has been based on sheath acceleration processes [3–5], a number of novel mechanisms have been at the centre of recent theoretical and experimental activities. Experiments in ion acceleration employ both ultrashort (10 s of fs) lasers systems (typically based on solid state, Ti:Sa technology) and high energy, picosecond laser systems (typically Nd:glass), with some experiments using CO_2 laser systems. This lecture will provide an overview of the main acceleration mechanisms and the underlying physical principles, as well as a brief review of the state of the art and recent developments in the field. More extensive surveys are provided in [6–8].

M. Borghesi (✉)
Centre for Plasma Physics, Queen's University Belfast, Belfast BT7 1NN, UK
e-mail: m.borghesi@qub.ac.uk

© Springer Nature Switzerland AG 2019

L. A. Gizzi et al. (eds.), *Laser-Driven Sources of High Energy Particles and Radiation*, Springer Proceedings in Physics 231, https://doi.org/10.1007/978-3-030-25850-4_7

7.2 Sheath Acceleration

This is the acceleration mechanism active in most experiments carried out so far, and it was proposed [4] as an interpretative framework of the multi-MeV proton observations reported in [2], obtained on the NOVA Petawatt laser at LLNL (the name Target Normal Sheath Acceleration, TNSA, is generally used).

Acceleration through this mechanism employs thin foils (typically from a few μm to tens of μm thickness), which are irradiated by an intense laser pulse. In the intensity regime of relevance (as a guideline, $I\lambda^2 > 10^{18}$ W/cm^{2}μm^2), the laser pulse can couple efficiently energy into relativistic electrons, mainly through ponderomotive processes (e.g. JxB mechanism [9]). The average energy of the electrons is typically of MeV order, e.g. their collisional range is much larger than the foil thickness, so that they can propagate to the rear of the target, and drive the acceleration of ions from surface layers via the space–charge field established as they try to move away from the target. While a limited number of energetic electrons will effectively leave the target, most of the hot electrons will be backheld within the target volume by the space charge, and will form a sheath extending by approximately a Debye length λ_D from the initially unperturbed rear surface. According to the model developed in [2], the initial accelerating field will be given by

$$E(0) = \frac{KT_h}{e\lambda_D} = \sqrt{\frac{n_h KT_h}{e\lambda_D}} \tag{7.1}$$

where n_h and T_h are density and temperature of the hot electrons, which for typical values at $I\lambda^2 \sim 10^{19}$ W/cm^{2}μm^2, i.e. $\lambda_d \sim 1$ μm and $T_h \sim 1$ MeV, gives field amplitudes of order TV/m. Under the right combination of target thickness and pulse duration, the hot electrons recirculate through the target during the ion acceleration process, which can lead to an enhancement of the ion energy [10]. TNSA from the front surface has normally reduced efficiency due to the presence of a preplasma, although symmetric acceleration from front and rear has indeed been observed in ultra-high contrast interactions with moderate intensity ultrashort pulses, where front preplasma formation is effectively suppressed [11].

While TNSA can in principle accelerate any ion species present in surface layers, in most experimental setting this results in preferential acceleration of light ions (protons, Carbon and Oxygen ions) from contaminant layers rather than ions from the target bulk. Protons, with the highest charge to mass ratio, are therefore the dominant component of TNSA ion beams, unless the target is suitably treated prior to the laser irradiation to remove the contaminants [12].

According to (7.1) the field can be large enough to accelerate ions to multi-MeV energies, which have indeed been observed in a very large number of experiments. The energy spectra of the ion beams observed are broadband (e.g. see Fig. 7.1), typically with an exponential profile, up to a high energy cut-off, which is the quantity normally used to compare different experiments and determine experimental scaling laws for the acceleration process.

Fig. 7.1 Proton spectrum
and conversion efficiency as
reported in Snavely et al.
2000 experiment in [1]

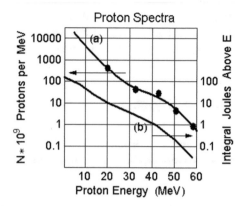

The highest TNSA energies reported are of the order of 85 MeV, obtained with
large PW systems, and available data (e.g. see Fig. 7.1) generally shows that, at
equal intensities, longer pulses (of ~ps duration) containing more energy generally
accelerate ions more efficiently than 10 s of fs pulses. Using state of the art fs systems
has however recently allowed increasing the energies of accelerated protons up to a
reported 40 MeV [13], obtained with only a few J of laser energy on target.

The properties of the beams accelerated via TNSA are quite different from those
of conventional RF beams, to which they are superior under several aspects. The
beams are characterized by ultralow transverse emittance (as low as 0.004 mm-
mrad, according to the estimate given in [14]), and by ultrashort (~ps) duration at the
source. The beams are bright, with 10^{11}–10^{13} protons per shot with energies >MeV,
corresponding to currents in the kA range if co-moving electrons are removed. However, the number of protons at the high-energy end of the spectrum (i.e. the energies
plotted in Fig. 7.2) can be as low as 10^7–10^8 particles/MeV/sr, (e.g. see [15] for a discussion related to recently published data)—with a divergence of a few degrees this

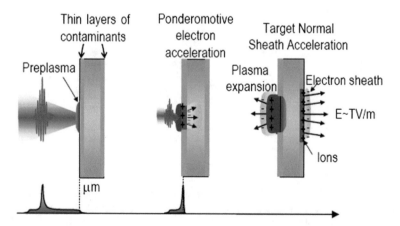

Fig. 7.2 Schematics of sequential stages of target normal sheath acceleration (from [110])

gives ~10^6–10^7 particles/MeV. Drawbacks, as compared with conventional accelerator beams, are the larger divergence (up to 10 s of degrees, and energy dependent) and, as mentioned earlier, the broad spectrum.

7.2.1 TNSA Scaling and Optimization

Increasing the laser intensity on target should generally lead to an increase of the cut-off energies of TNSA spectra, as shown in Fig. 7.3. However there is still debate on what is the most appropriate scaling for ion energies as a function of irradiance, and is also clear that, in addition to the role of pulse energy, several secondary factors (e.g., such as prepulse energy and duration, target thickness) also affect the maximum energy measurable.

Parametric investigations of the dependence of E_{max} on laser pulse irradiance, duration, energy and fluence have been reported (e.g. [16–19]). Two main classes of approaches have been developed to describe analytically the TNSA process with the aim of matching current results and predict performance at higher intensities. A first approach considers ions and hot electrons as an expanding plasma, described with fluid models [4, 5, 20] as an extension of the classical case of a plasma expanding

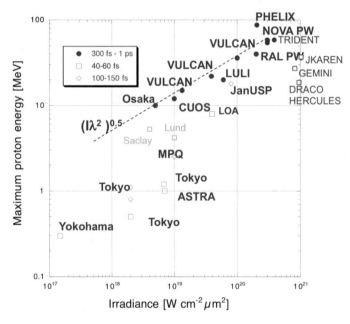

Fig. 7.3 Survey of TNSA cut-off energies measured in experiments, plotted vs irradiance and labelled according to pulse duration. For references for the specific data points, see [6, 13]. Points labelled J-Karen, DRACO and Trident refer respectively to [13, 16, 17]. The point labelled PHELIX refers to the work reported in [111]

into vacuum, driven by the ambipolar electric field generated in a narrow layer at the front of the plasma cloud. Simplest models are isothermal, and require that the acceleration time is artificially constrained [18, 19], while more realistic adiabatic models, accounting for the finite energy of the hot electrons, have also been developed [20].

A different class of models assumes that the most energetic ions are accelerated as test particles in a *static* sheath field, unperturbed by their acceleration. These static models rely on an accurate description of the sheath field based on realistic assumptions on the fast electron distribution. For example, in [21], a spatial truncation of the electric potential in the sheath is introduced, and used to develop a model for the maximum ion energy as a function of the relevant laser parameters (energy and intensity). Scalings for the ion energy based on this model appear to match a large fraction of experimental results so far [22], and can be used as a predictive tool for future performance. Taking 200 MeV H$^+$ energy as a benchmark, predictions of the intensity requirement for reaching this cut-off value based on the two different approaches discussed above give intensities of mid 10^{21} W/cm^2 for ~ps pulses [21], and ~10^{22} W/cm^2 for 10 s of fs pulses [22].

Experimentally, several approaches have been developed to improve TNSA efficiency by acting on the characteristics of the hot electron population driving the acceleration, through modifications of the target design [8]. According to (7.2) the accelerating field can be modified either by increasing the electron density or the temperature. The use of the so-called *mass limited targets*, aims to reduce the transverse size of the accelerating foils and concentrating the electrons with in a smaller volume so that the density is increased during the acceleration process. This approach was first demonstrated in [23] where reduction of the foil down to 20 μm × 20 μm resulted in a 3-fold protons energy increase with respect to a large mm-size foil, jointly to a sizeable increase in conversion efficiency. A further class of experiments aims to optimize laser energy absorption into hot electrons by structuring the target: an example of this approach is reported in [24], where foils coated with microspheres (diameter ~λ/2) on the irradiated surface showed, compared to uncoated foils, a clear improvement in the energy cut-off of the spectrum of the accelerated protons.

Targets with special microstructuring and/or shaping of the rear side have been also used for spectral and spatial manipulation of the proton beam. For example, target shaping to manipulate field configuration has been used for beam focusing (e.g., see [25, 26]) and the highly transient nature of the TNSA field has been used for dynamic focusing with chromatic capability using a two-beam configuration [27].

A recently proposed approach employs what is a by-product of intense laser-target interactions (i.e. the generation of large amplitude electromagnetic pulses, EMP, through a process akin to ultrafast electric dipole emission), to control the properties of TNSA-accelerated ions. This is done by attaching a coiled metallic wire behind the target, and synchronizing the propagation of TNSA protons of a given energy with the EMP propagating along the coil. The strong electric field associated to the EMP can act on the protons by constraining their divergence but also by re-accelerating the protons. In proof-of-principle experiments, doubling of the energies of TNSA protons has been achieved through this process, jointly to the production of a highly collimated, narrow band proton beamlet [28].

For a review of other approaches, including the use of foam layers, controlled pre-plasmas, or double pulses, see [8].

7.3 Beyond TNSA: Emerging Mechanisms

While experimental activity has focused until recently on the study of TNSA beams, other mechanisms have also attracted a significant amount of theoretical and experimental attention, and some of these mechanisms are briefly discussed in the following sections. We refer the reader to [8] for a more thorough discussion. Please note that, although conceptually different, these mechanisms can coexist during an interaction, and in many experimental settings (particularly when employing ultrathin foils), ion acceleration will proceed through hybrid processes combining elements of different acceleration mechanisms, including TNSA.

7.3.1 Radiation Pressure Acceleration

It is well known, already from Maxwell's e.m. theory, that electromagnetic (EM) waves carry momentum, and this momentum may be delivered to a non-transparent (either absorbing or reflecting) medium irradiated by the EM wave.

In a classical approach, the momentum p carried by the wave per unit volume is given by $p = w/c$, where w is the e.m. energy density and c the speed of light. If the light strikes a surface A, it will apply a force to the surface, which is equal and opposite to the rate of change of the momentum of the wave. If the light is completely absorbed by the surface, the momentum change per unit time is given by the e.m. momentum contained within a volume cA, the force applied to the surface is $F = pcA$, and the pressure is $P_{rad} = F/A = w = I/c$, where I is the intensity of light.

In case of light being fully reflected from the surface, the change of momentum is double than in the total absorption case, as the momentum of the light is reversed after reflection, and the force applied to the surface would therefore be $F = 2pcA$ from which $P_{rad} = 2I/c$.

For an intense laser pulse, such pressure can be enormous: for example for $I = 10^{20}$ W cm^{-2} (as achievable nowadays with state of the art lasers), the pressure $2I/c$ is of the order of 60 Gbar. Such pressure can strongly alter the dynamics of laser-plasma interaction, and can be used to accelerate and propel forward particles in the plasma.

Hole Boring. If an intense laser pulse irradiates a dense plasma, the Radiation Pressure is coupled to the electrons via the ponderomotive force

$$F_{p0} = -\frac{e^2}{4m_e\omega^2}\nabla E_0^2 \tag{7.2}$$

Note that the expression above is obtained, for linearly polarized light, as a cycle average of the expression:

$$F_p(t) = -\frac{e^2}{2m_e\omega^2}\nabla E_0^2 cos^2(\omega t) = -\frac{e^2}{4m_e\omega^2}\nabla E_0^2(1 + cos(2\omega t)) \tag{7.3}$$

(e.g. see [29]). The instantaneous ponderomotive force is therefore composed of two terms, the steady ponderomotive force F_{p0} and a term oscillating at twice the frequency of the radiation. This oscillating term is responsible for electron heating through the so-called *Ponderomotive (or JxB) heating* [9]. As we will discuss later heating electrons to high temperature is generally detrimental for radiation pressure acceleration. However, if one employs circularly polarized light at normal incidence, the oscillating component of the force is not present, and $F_p(t) = F_{p0}$. For this reason most theoretical and numerical investigations of RPA (as well as some of the experiments) employ circularly polarized pulses.

At the surface of the overdense plasma the electrons are pushed inward by F_{p0}, leaving a charge separation layer and creating an electrostatic, backholding field that in turn acts on the ions and leads to their acceleration. This dynamics leads to the process called *hole boring* [30], i.e. a dynamic deformation of the plasma density profile which allows the laser to penetrate into the plasma. During this process the ions are compressed into a front which is pushed forward by the ponderomotive force, and the name *hole boring acceleration* is typically used to describe the ion dynamics in this process. The recession velocity of the plasma surface (i.e. the hole boring velocity u_{HB}) can be obtained from momentum balance considerations (and assuming the velocity of the ions is non relativistic) as [30]:

$$\frac{u}{c} = \left[\frac{n_{cr}}{2n_e}\frac{Zm_e}{m_i}\frac{I\lambda^2}{1.37 10^{18}}\right]^{\frac{1}{2}}$$

where n_e and n_{cr} are the electron plasma density and the critical density respectively. A complete, fully relativistic derivation of the hole boring velocity leads to a more complex dependence (e.g. see [31]), however even from this simple, 1D description some interesting features emerge:

(1) The energy of the ions in the hole boring front $E_i = mu^2/2$) is directly proportional to the intensity of the laser.
(2) E_i has an inverse proportionality to the plasma density (and consequently to the mass density $\rho \sim m_i n_e/Z$)), which suggests that the ion energy obtainable through this process can be optimized by choosing plasmas or targets of suitably low density (provided they are not transparent to the radiation).

A more detailed understanding of the dynamics of ion acceleration during hole boring can be had from the cartoon provided in Fig. 7.4 (see also [32]), which

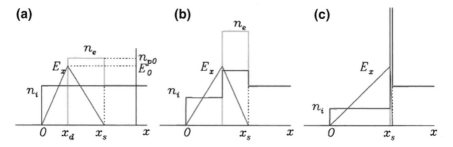

Fig. 7.4 Sketch of electron density n_e, ion density n_i and longitudinal electric field E_x at sequential stages of the hole boring process during intense irradiation of an overdense plasma

assumes an initial step-density profile with $n_e > n_{cr}$. The ponderomotive force of the laser quickly pushes forward the electrons, which pile up, and leave behind a layer of non neutralised ions. The charge distribution creates an electrostatic field E_x, which accelerate the ions. The ions located between x_d and x_s experience an electrostatic field which decrease with distance, meaning that the ions at the back of this layer are accelerated more than those at the front, which results in bunching of the ions, ultimately leading to a ion spike at the point where the electric field was initially zero, as shown in Fig. 7.4c, and to an average velocity u_{HB}. Simulations [32] indicate that during this process, collapse of the electron equilibrium leads to a non-linear phase in which a narrow bunch of fast ions can be accelerated at $u \sim 2 u_{HB}$.

Experimental data on HB acceleration using solid density foils is limited so far, mainly because the energies of HB-accelerated ions are low due to the inverse dependence on target density and therefore ions from TNSA processes dominate experimental spectra. Most interesting for acceleration via this method are media of low density, where the ρ^{-1} dependence can be exploited. Results which have highlighted HB-accelerated ions have been obtained employing CO_2 lasers ($\lambda \sim 10\ \mu m$). Due to the long wavelength, the critical density is reduced by ~100 times with respect to the more broadly used solid state lasers (e.g. Nd:YAG or Ti:Sa systems). Therefore even gas jet targets can be overdense for CO_2 laser pulses. For example, in an experiment employing CO_2 laser pulses and a gas jet [33], proton spectra obtained through this method showed a clean monoenergetic signal at ~1 MeV which was broadly consistent with hole boring acceleration in the conditions of the experiment. Although the ion energy observed so far through HB is modest, the dependence on both target density and laser intensity allows designing suitable acceleration scenarios for acceleration to 100 s MeV/nucleon with next generation laser facilities, as suggested in [31, 34].

An acceleration scheme related to HB, but conceptually different, is the so-called **Shock Acceleration**, first proposed in [35]. In this scheme, the light pressure applied at the front surface of the target, acts as the source of a strong, collisionless electrostatic shock propagating towards the bulk of the plasma. Acceleration arises as ions present in the bulk of the target are reflected from the shock front to twice the shock velocity, in a similar fashion to acceleration scenarios thought to take place in astrophysical contexts. Results also obtained with a CO_2 laser have been explained with this mechanism, namely monoenergetic acceleration of protons up to 22 MeV from the interaction with hydrogen gas jets at intensities in the 10^{16}–10^{17} W cm^2 regime [36].

Light Sail. Light Sail is the name currently used to define a different regime of Radiation Pressure Acceleration where the irradiated target is thin enough that the Hole boring process reaches the rear of the target before the end of the laser pulse [8, 37–40]. In such a case, the laser pulse is able to further accelerate ions to higher energies since the ions are not screened by a background plasma anymore. In an ideal situation, the irradiated region of the target is detached and pushed forward under the effect of the intense Radiation Pressure of the laser pulse, as sketched in Fig. 7.5.

An estimate of the dependence of the ion energy in this regime from the relevant parameters can be obtained from simple considerations. Neglecting absorption, the force applied by the Radiation pressure to an area A of the targets is given by

$$F = (1 + R)A\frac{I}{c}$$

The momentum acquired by the initially stationary target will be equal to the product of this force by the pulse duration τ, and the mass contained in the irradiated area of the target is given by $m_i n_i Ad = \rho Ad$. Therefore one has that the final velocity acquired by the target will be

$$u_i = (1 + R)\frac{1}{\rho d}\frac{I}{c}\tau$$

which gives the dependence

$$E_i \sim \left(\frac{I\tau}{\rho d}\right)^2$$

A few interesting considerations can be made on the basis of this expression:

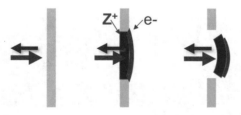

Fig. 7.5 Schematic representation of Light Sail acceleration. Reflection of light at the target surface causes a pressure on the electrons driving hole boring through the target. The target is thin enough that the laser keeps driving the irradiated portion of the target after the hole boring front has reached its rear

(1) The scaling of ion energy with laser intensity is much faster than in HB (and TNSA), which is promising for acceleration to high energy.
(2) Due to the product $I\tau$, the dependence is on fluence (Energy per area) rather than on intensity only.
(3) There is an inverse dependence on the areal density $\eta = \rho d$ rather than on the target density which points to the advantage of using as thin targets as possible, provided that they are not so thin to be transparent to the radiation. An optimum thickness emerges if one takes transparency into consideration, as shown in [40].

The scheme has been widely investigated via numerical simulations, which, while confirming the general features above, have highlighted a complex dynamics, in which charge separation between the electron and ions is maintained by the ponderomotive force of the laser as the compressed electron ion layer becomes detached from the target.

Figure 7.6 shows the implementation of such a scheme in 2D Particle in Cell simulations. The sequential frames show ion density plots at different stage of the process, indicating how a compressed ion layer is detached and pushed forward by the radiation (the laser is still active in the last frame shown). The red curve in Fig. 7.7 is an ion energy spectrum from the same simulation, characterized by a spectral peak at very high energies (>GeV). A quasi-monoenergetic spectrum arises as all ions in the compressed layer share the same acceleration history (differently from TNSA spectra, which are typically very broad and exhibit an exponential decay at the high energy end).

For LS to be effective, it is essential that the radiation pressure is strong enough to overcome detrimental effects related to electron heating, such as foil disassembly under the thermal pressure of hot electrons, or debunching of the compressed foil, and to dominate over TNSA. As discussed before, the use of circularly polarized light can in principle reduce these effects, although some electron heating is unavoidable.

Recent work has, however, highlighted hybrid RPA-TNSA regimes using linearly polarized pulses, where, under appropriate conditions, RPA features dominate the ion spectra [41], and has investigated the more complex dynamics associated with RPA of multispecies targets.

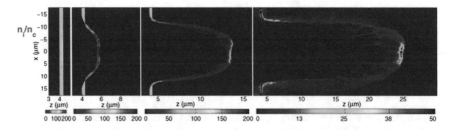

Fig. 7.6 Snapshots of 2D PIC simulations of Light Sail acceleration of protons from a thin hydrogen target. The pulse intensity employed in the simulation was circularly polarised and had intensity ~5 10^{22} W/cm^2 See [39] for further information

Fig. 7.7 Energy spectra of protons accelerated by Light Sail in 2D PIC simulations. The red curve is from the simulation shown in Fig. 7.6. The green profile refers to a case in which the acceleration process is unstable (see [39])

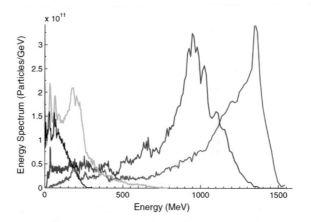

Experiments employing ultrathin foils have recently started to show signatures of RPA acceleration processes, namely effective acceleration of bulk species, a strong polarization dependence and the emergence of spectral peaks (see, e.g., [42], reporting recent results from the GEMINI 40 fs laser system, at intensities of ~5 10^{20} W/cm^2—see Fig. 7.8).

Features consistent with hybrid TNSA-RPA regimes have been observed in experiments employing ultrathin foils and sub-ps, PW-class laser pulses [43]. Record proton

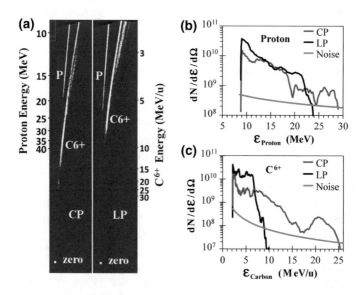

Fig. 7.8 Ion spectra from irradiation of ultrathin foils on the GEMINI laser (from [42]) (**a**) Example Thomson parabola traces for circularly (CP) and linearly (LP) polarised laser pulses irradiating 10-nm amorphous carbon targets. The corresponding CP (red) and LP (black) proton (**b**) and C$_6$ spectra (**c**) are also shown. The noise level is also plotted

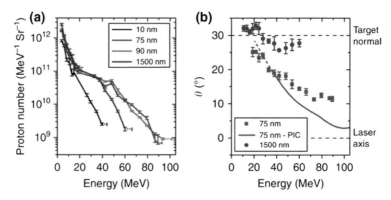

Fig. 7.9 Measurements of proton beam spectrum and direction for thin CH foils irradiated by the VULCAN PW laser (from [44]). **a** Example proton energy spectra, for given foil thickness, ℓ. The highest proton energies are observed for 75–90 nm thick targets, where the laser pulse undergoes relativistically induced transparency near the peak of the pulse. **b** Measured angle of the centre of the proton beam, θ, with respect to the laser axis (in the plane of the incident laser beam), as a function of energy, for $\ell = 75$ nm (red) and 1.5 μm (blue). An example PIC simulation result for $\ell = 75$ nm (red curve) is included for comparison. The dashed lines mark the target normal and laser axis, for ease of reference

energies (approaching 100 MeV) have been recently reported in an experiment carried out on VULCAN in which a hybrid TNSA-RPA regime was optimized and enhanced by relativistic transparency processes (see next section and Fig. 7.9) [44].

7.3.2 Relativistic Transparency Regimes

Acceleration regimes in which the target becomes relativistically transparent to the laser pulse are also of interest, and have been explored in a number of experiments [45–48]. In these investigations the target areal density is chosen so that the target is quickly heated by the laser pulse, and the density decreases below the relativistically corrected critical density near the peak of the pulse. In this regime the interaction leads to volumetric heating of the target electrons, and to a consequent enhancement of the field accelerating the ions. In the Break Out Afterburner scenario proposed by the Los Alamos group [49], non-linear processes lead to growth of electromagnetic instabilities, which further enhances energy coupling into the ions.

Experimental spectra obtained in this regime are generally broadband, with particle numbers decreasing to a high energy plateau and show efficient acceleration of the bulk components of the target. The mechanisms of this interaction regime are rather complex, and a number of recent investigations have been devoted to the electron and ion dynamics during the interaction, as well as their dependence on laser polarization [48, 50]. Under appropriate conditions, occurrence of transparency near the peak of the laser pulse can lead to an enhancement of the energy of the accelerated ions (e.g. as in [44]), as well as to distinctive angular emission patterns [50, 51].

7.4 Applications of Laser-Driven Ions

In the following section some of the main applications of laser-driven ion beams will be briefly reviewed. We will discuss some applications which can already be implemented with current beam parameters (proton radiography, warm dense matter generation, pulsed radiobiology, neutron beam production) as well as more speculative applications, which require an uplift in performance of laser-accelerators (Fast Ignition of Thermonuclear Fusion and particle therapy of cancer), and may be facilitated by the next generation of ultra-intense laser systems.

7.4.1 Proton Radiography/Deflectometry

The unique properties of protons from high intensity laser-matter interactions, particularly in terms of spatial quality and temporal duration, have opened up a totally new area of application of proton probing/proton radiography. Several experiments have been carried out in which laser-driven proton beams have been employed as a backlighter for static and dynamic target assemblies, typically a secondary target irradiated by a separate laser pulse. In light of the high laminarity of laser-driven proton beams, the protons emitted from a laser-irradiated foil can be thought of as emitted from a virtual, point-like source located in front of the target [52]. A point-projection imaging scheme is therefore automatically achieved with magnification set by the geometrical distances at play. Density variations in the target probed can be detected via modifications of the proton beam density cross section, caused by differential stopping of the ions, or by scattering. Similarly, electric or magnetic fields in the sample region can be revealed by the proton deflection and the associated modifications in the proton density pattern.

Backlighting with laser-driven protons has intrinsically high spatial resolution, which, for negligible scattering in the sample investigated, is determined by the size d of the virtual proton source and the width δs of the point spread function of the detector, offering the possibility of resolving details with spatial dimensions of a few μm. However, resolution degradation due to multiple scattering affects this technique when the sample probed is not very thin, effectively increasing the size of the backlighting proton source. Multilayer detector arrangements employing RCFs or CR39 layers offer the possibility of energy-resolved measurements despite the beam's broad energy spectrum. Energy dispersion provides the technique with an intrinsic multi-frame capability. In fact, since the sample to be probed is situated at a finite distance from the source, protons with different energies reach it at different times. As the detector performs spectral selection, each RCF layer contains, in first approximation, information pertaining to a particular time [53]. Depending on the experimental conditions, 2-D frames spanning up to 100 ps can be obtained in a single shot. The ultimate limit of the temporal resolution is given by the duration of the proton burst τ at the source, which is of the order of the laser pulse duration. However other effects are also important: the finite energy resolution of the detector layers and the finite transit time of the protons through the region where the fields are present normally limit the resolution to a few ps.

Several radiographic applications of laser-produced protons have been reported to date.

Density diagnosis via proton radiography has potential application in Inertial Confinement fusion. A preliminary test studying the compression of empty CH shells under multi-beam isotropic irradiation at the moderate irradiance of 10^{13} W/cm^2 was carried out at the Rutherford Appleton Laboratory, where radiographs of the target at various stages of compression were obtained [54]. Experiments aiming to use protons as a shock diagnostic in laser-irradiated dense targets have also been carried out [55], although the aforementioned scattering degradation effects have limited these experiments to low-density foam targets at currently available proton energies. Projection radiography of static objects (where high-resolution images are imprinted on a suitable detector by either scattering or stopping processes) has also been explored in a number of experiments, employing point projection [52, 56] or contact radiography [57, 58] schemes.

The most successful applications to date of proton backlighting are related to implementations of this technique aimed to detect electric and magnetic fields in plasmas [53], via the deflections undergone by the protons. This has made possible obtaining for the first time direct information on electric fields arising through a number of laser-plasma interaction processes, and has provided a powerful tool for magnetic field measurements. In this way novel and unique information has been obtained on a broad range of plasma phenomena. The high temporal resolution is here fundamental in allowing the detection of highly transient fields following short pulse interactions.

Two main arrangements have been explored: in proton imaging (i.e. simple back-lighting projection of the sample), the deflections cause local modulations in the proton density n_p across the proton beam cross section, which, under simplified assumptions, can yield line-averaged values of the fields [59]. In a proton deflectom-etry arrangement, thin meshes are inserted in the beam between the proton source and the object as "markers" of the different parts of the proton beam cross section [60]. The meshes impress a modulation pattern in the beam before propagating through the electric field configuration to be probed. The beam is in this way effectively divided in a series of beamlets, and their deflection can be obtained directly from the distortion of the impressed pattern.

A general analysis method, applicable to both arrangements, consists of using particle tracing codes to follow the propagation of the protons through a given three-dimensional field structure, which can be modified iteratively until the computa-tional proton profile reproduces the experimental ones. State-of-the-art tracers allow

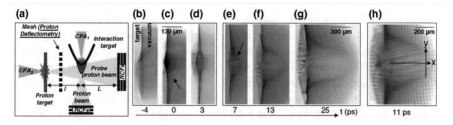

Fig. 7.10. a Set-up of a proton radiography experiment investigating sheath expansion from the rear of a laser irradiated target. **b–g** Typical proton imaging data at probing times −4, 0, 3, 7, 13, and 25 ps, respectively, and **h** proton deflectometry data. The magnification was 30 in (**b**)–(**d**) and 15 in (**e**)–(**h**). The scales refer to the interaction target plane. From [61]

realistic simulations including experimental proton spectrum and emission geometry, as well as detector response.

Data obtained with this technique are shown in Fig. 7.10. In this case, the protons are used to probe the rear of a foil following ultraintense irradiation of the front of the foil (the foil is curved to allow access to the region near the target surface where the fields are more intense) [61]. The probe proton pattern is modified by the fields appearing at the target rear as a consequence of the interaction, and the technique effectively allows spatially and temporally resolved mapping of the electrostatic fields associated to TNSA acceleration of protons from the foil. The data confirmed the existence of a large short-lived (flash) field at the target surface temporally coincident with the irradiation ($\sim 10^{11}$ V/m), and allowed following the consequent expansion of the accelerated ion front, via detection of the field at the front. Both the initial sheath field and the ion front show a characteristic bell shape. The frames (b–g) are taken in a single shot and highlight clearly the capability of this diagnostic to produce a "movie" made-up of discrete frames. A modified arrangement enabling detection of field fronts propagating at relativistic speeds was proposed by Quinn et al. which has allowed resolving the propagation of an electric field pulse along the surface of a laser irradiated target, with v ~ c [62].

Proton radiography is now a well-established diagnostic technique employed by many groups worldwide, for the investigation of a broad range of plasma phenomena (see [8] for a more extensive literature survey). Active areas of investigation using radiography techniques with TNSA protons include the dynamics of large-scale magnetic fields [63, 64], and the study of collisionless shock physics [65–67] of astrophysical relevance.

7.4.2   Warm Dense Matter Studies

Laser-driven ions have also found application in a number of experiments aimed to heat up solid density matter via isochoric heating, and create so-called Warm Dense Matter (WDM) states (i.e. matter at 1–10 times solid density and temperatures up to 100 eV) [68] of broad relevance to material, geophysical and planetary studies. The high-energy flux and short temporal duration of laser-generated proton beams are crucial parameters for this class of applications. The creation of so-called Warm Dense Matter (WDM) states can be achieved by several other means. However, when studying fundamental properties of WDM, such as equation of state or opacity, it is desirable to generate large volumes of uniformly heated material; ion beams, which can heat the material in depth, are in principle better suited to this purpose than other methods available, such as shock heating or x-ray or electron heating.

Heating of solid density material with ions can be achieved with accelerator-based or electrical-pulsed ion sources. However, the relatively long durations of ion pulses from these sources (1–10 ns) means that the materials undergo significant hydrodynamic expansion already during the heating period. On the contrary, laser-generated proton beams, emitted in ps bursts, provide a means of very rapid heating, on a timescale shorter than the hydrodynamic timescale. By minimizing the distance between the ion source and the sample to be heated, it is possible to limit the heating time to 10 s of ps. The target then stays at near-solid density for 10–100 s ps before significant expansion occur, and the WDM properties can be investigated within this temporal window. (Fig. 7.11)

The first demonstration of laser-driven proton heating was obtained by Patel et al. [25]. In this experiment a 10 J pulse from the 100 fs JanUSP laser at LLNL was focused onto an Al foil producing a 100–200 mJ proton beam. A second 10 μm thick Al foil was placed in the path of the proton beam a distance of 250 μm from the first. Target heating was monitored via time-resolved rear surface emission, providing a measurement of the initial temperature of the heated Al (~4 eV). A focused proton beam, produced from a spherically-shaped target, was seen to heat a smaller region to a significantly higher temperature, approximately 23 eV. Subsequent experiments have mostly foregone ballistic focusing, easing set-up constraints and placing the emphasis on the characterization of the heated sample. In these experiments, a naturally diverging laser-driven proton beam is employed to heat a secondary target placed very close (typically hundreds of microns) to the ion source. The properties of the Warm Dense Matter so produced are then investigated with a number of diagnostics, either passive or in pump-probe configurations, combined to self-consistent modeling of sample heating and expansion, which also requires a full characterization of the proton beam parameters, down to the low energy end of the spectrum. In particular, pump-probe arrangements have led to totally novel information on the transition phase between cold solid and plasmas in isochorically heated targets, as observed for example in [69–71].

The short burst duration of laser-driven protons has also been recently exploited in investigations of the transient dynamics of proton-irradiated samples, by observing variations in the optical opacity of transparent materials on ps timescales [72, 73].

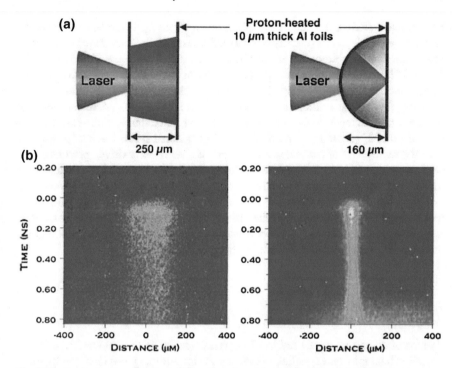

Fig. 7.11. Proton heating data from [25]. **a** Experimental setup for flat and focusing target geometries. Each target consisted of a flat or hemispherical 10 μm thick Al target irradiated by the laser, and a flat 10 μm thick Al foil to be heated by the protons. **b** Corresponding streak camera images showing space- and time-resolved thermal emission at 570 nm from the rear side of the proton-heated foil. The streak camera images an 800 μm spatial region with a 1 ns temporal window

7.4.3 Radiobiology

Several groups have initiated experimental activities in which laser-driven proton beams have been used to irradiate cellular samples, in order to investigate the biological response of the cells to the protons. This work has been motivated by the proposed future use of laser-driven ions in cancer therapy (see below) but also by the possibility of accessing unexplored regimes of radiobiology at ultra-high dose rate, thanks to the ultrashort duration of the ion bursts. In a typical arrangement, doses of up to a few Gy can be delivered to the cells in short bursts of ∼ns duration. In some experiments [74, 75] the dose is fractionated and the average dose rate is comparable to the one used in irradiations with conventional RF accelerators ($\sim 0.1\,\mathrm{Gy\,s^{-1}}$). In single-shot irradiations, on-cell dose rates of the order of $10^9\,\mathrm{Gy\,s^{-1}}$ have been estimated [76–78]. In view of this enormous difference in dose rate, and of its possible effects on the cell response, experiments have initially focused on assessing the biological effect of laser-driven ions with respect to conventionally accelerated ion beams and other reference radiation. To this aim, the Relative Biological

Effectiveness (RBE), defined as the ratio D_X/D_p where D_x is a reference dose of a standard radiation source (usually X-rays) and D_p is the laser accelerated proton dose producing the same biological effect, has been measured in several experiments [74, 76, 77, 79], which have all shown results broadly in line with results obtained at conventional dose rates. These experiments have employed traditional cellular assays, assessing the damage inflicted to the cells' DNA or the fraction of cells surviving the irradiation. More recently, experiments investigating sublethal effects of laser-driven protons [80, 81] (i.e. long-term effects on cells irradiated at low levels of dose) have reported indications of a less negative impact of laser of laser-driven protons on the cells compared to irradiations with *conventional* proton beams at lower dose rate. This is encouraging for possible future therapeutic use of laser-driven protons as, in a clinical context, would translate in lower damage to healthy cells traversed by the ions on their way to a tumour (as recently observed, for example, in short pulse electron (FLASH) irradiations) [82]. For a more detailed discussion of the radiobiology experiments carried out so far, we refer the reader to [83].

7.4.4 Cancer Therapy

Protons or carbon ions are used in ~70 centres worldwide to treat cancer [84]. The use of ion beams in cancer radiotherapy exploits the advantageous energy deposition properties of ions as compared to more commonly used X-rays, allowing for a better localization of the dose to the target tumor and a reduction of harmful effects to the healthy tissues surrounding it [85]. The proton energy window of therapeutic interest ranges between 60 and 250 MeV, depending on the location of the tumor (the required Carbon ion range extends up to 350 MeV/nucleon). Most of the facilities use protons, with a minor number using carbon ions which are more effective for radioresistant and hypoxic tumors [86].

The potential use of laser-driven protons for future cancer therapy was originally proposed in a number of papers, which argued potential advantages in compactness and cost compared to conventional systems based on RF accelerators [87–89]. While these arguments need to be reassessed in view of the recent progress in compact proton therapy systems, interest in a possible laser-driven approach remains significant, with a number of projects currently active which aim to explore and develop the potential of laser-driven ion sources towards cancer therapy applications (e.g. see a review in [90]).

It is recognized that there are significant challenges ahead before laser-driven ion beams capable of meeting therapeutic specifications, both in terms of energy, repetition rate and general reliability, may become available. However, several authors have started to consider how a laser-driven treatment could be delivered, for example designing reduced size magnetic delivery systems (gantries) [91] or developing treatment plans tailored to the characteristics of TNSA beams [92, 93]. Another significant development in this area is the future availability of laser-driven proton beamlines with controlled output parameters, which will allow significant progress in preclinical work (e.g. see [94]). For a more extensive assessment of the prospects and interest of laser-driven ions towards cancer therapy, we refer the reader to [95].

7.4.5 Neutron Generation

The interaction of laser-driven high-energy ions with secondary targets can intiate nuclear reactions of various type, presenting the opportunity of carrying out nuclear physics experiments in laser laboratories rather than in accelerator facilities [96, 97], and to apply the products of the reaction processes in several areas. In particular, nuclear reactions driven by laser-accelerated proton or deuteron beams are of interest for the production of beams of high energy neutrons, which can be initiated by directing the ion beam onto a secondary target, i.e. in a "pitcher-catcher" configuration. A number of experiments employing laser-accelerated protons and deuterons has reported the acceleration of beams of *fast* neutrons (with energies from MeV to 10 s of MeV) [98–102], which typically display a clear directionality/anisotropy along the propagation axis of the ion beam.

Laser-driven neutron sources have potential advantages over conventional reactor- or accelerator-based sources in term of cost, compactness, and short duration for applications such as neutron resonance spectroscopy, fast neutron radiography and material testing [103]. Recently, the moderation of MeV laser-produced neutrons to epithermal energies (eV–100 keV) has been demonstrated [104], which can in principle open up a broader range of applications in neutron science, provided sufficient neutron fluxes can be obtained on next generation laser-systems.

7.4.6 Proton Fast Ignition

The use of laser-driven ions as a trigger to start ignition in a compressed fusion fuel pellet was first proposed by Roth et al. [105], as an alternative to the electron-driven Fast Ignitor concept [106]. The concept proposed in [105] envisaged coupling laser-driven protons to the interior of an indirect drive assembly and taking advantage of the energy deposition profile of the protons, their short duration and their focusability. Detailed analysis of this scheme has been carried out, e.g. in [107, 108], using the parameters of TNSA ions. The problem to overcome in this approach is not the required energy of the protons, which is in the 10–20 MeV range, but the very high number of particle required to heat the hot spot (which translates in laser pulses energies up to ~100 kJ, considering reported laser-to-proton conversion efficiencies [108]). More recent work has considered ignition using RPA accelerated carbon ions or ions accelerated ponderomotively via Hole Boring acceleration. A recent survey of the relative merit and requirements of these approaches is provided in [109].

Acknowledgements The author acknowledges support from EPSRC (grants EP/K022415/1 and EP/J500094/1).

References

1. E.L. Clark et al., Phys. Rev. Lett. **84**, 670 (2000)
2. A. Maksimchuk et al., Phys. Rev. Lett. **84**, 4108 (2000)
3. R.A. Snavely et al., Phys. Rev. Lett. **85**, 2945 (2000)
4. S.C. Wilks et al., Phys. Plasmas **8**, 542 (2001)
5. P. Mora, Phys. Rev. Lett. **90**, 185002 (2003)
6. M. Borghesi et al., Fusion Sci. Tech. **49**, 412 (2006)
7. H. Daido, M. Nishiuchi, A.S. Pirozkhov, Rep. Prog. Phys. **75**, 056401 (2012)
8. A. Macchi, M. Borghesi, M. Passoni, Rev. Mod. Phys. **85**, 751 (2013)
9. W.L. Kruer, K. Estabrook, Phys. Fluids **28**, 430 (1985)
10. A.J. Mackinnon et al., Phys. Rev. Lett. **88**, 215006 (2002)
11. T. Ceccotti et al., Phys. Rev. Lett. **99**, 185002 (2007)
12. M. Hegelich et al., Phys. Rev. Lett. **89**, 085002 (2002)
13. K. Ogura et al., Opt. Lett. **37**, 2868 (2012)
14. T. Cowan et al., Phys. Rev. Lett. **92**, 204801 (2004)
15. A. Macchi et al., Plasma Phys. Control. Fusion **55**, 124020 (2013)
16. K. Zeil et al., New J. Phys. **12**, 045015 (2010)
17. K.A. Flippo et al., Rev. Sci. Instr. **79**, 10E534 (2008)
18. J. Fuchs et al., Nat. Phys. **2**, 48 (2006)
19. L. Robson et al., Nat. Phys. **3**, 58 (2007)
20. P. Mora, Phys. Rev. E **72**, 056401 (2005)
21. M. Passoni, M. Lontano, Phys. Rev. Lett. **101**, 115001 (2008)
22. M. Passoni, L. Bertagna, A. Zani, New J. Phys. **12**, 045012 (2010)
23. S. Buffechoux et al., Phys. Rev. Lett. **105**, 015005 (2010)
24. D. Margarone et al., Phys. Rev. Lett. **109**, 234801 (2012)
25. P.K. Patel et al., Phys. Rev. Lett. **91**, 125004 (2003)
26. S. Kar et al., Phys. Rev. Lett. **100**, 105004 (2011)
27. T. Toncian et al., Science **312**, 410 (2006)
28. S. Kar et al., Nat. Comm. **7**, 10792 (2016)
29. F.F. Chen, *Introduction to Plasma Physics and Controlled Fusion* (Plenum Press, 1984)
30. S.C. Wilks et al., Phys. Rev. Lett. **69**, 1385 (1992)
31. A.P.L. Robinson et al., Plasma Phys. Control. Fusion **51**, 024004 (2009)
32. A. Macchi et al., Phys. Rev. Lett. **94**, 165003 (2005)
33. C.A.J. Palmer et al., Phys. Rev. Lett. **106**, 014801 (2011)
34. A.P.L. Robinson et al., Plasma Phys. Control. Fusion **54**, 115001 (2012)
35. L. Silva et al., Phys. Rev. Lett. **92**, 015002 (2004)
36. D. Haberberger et al., Nat. Phys. **8**, 95 (2012)
37. T. Esirkepov et al., Phys. Rev. Lett. **92**, 175003 (2004)
38. A.P.L. Robinson et al., New J. Phys. **10**, 013021 (2008)
39. B. Qiao et al., Phys. Rev. Lett. **102**, 145002 (2009)
40. A. Macchi, S. Veghini, F. Pegoraro, Phys. Rev. Lett. **103**, 085003 (2009)
41. B. Qiao et al., Phys. Rev. Lett. **108**, 115002 (2012)
42. S. Kar et al., Phys. Rev. Lett. **109**, 185006 (2012)
43. C. Scullion et al., Phys. Rev. Lett. **119**, 054801 (2017)
44. A. Higginson et al., Nat. Comm. **9**, 724 (2018)
45. A. Henig et al., Phys. Rev. Lett. **103**, 045002 (2009)

46. D. Jung et al., New J. Phys. **15**, 023007 (2013)
47. D. Jung et al., Phys. Plasmas **20**, 083103 (2013)
48. B. Gonzalez-Izquierdo et al., Nat. Comm. **7**, 12891 (2016)
49. L. Yin et al., Phys. Rev. Lett. **107**, 045003 (2011)
50. B. Gonzalez-Izquierdo et al., Nat. Phys. **12**, 505 (2016)
51. H. Powell et al., New J. Phys. **17**, 103033 (2015)
52. M. Borghesi et al., Phys. Rev. Lett. **92**, 055003 (2004)
53. M. Borghesi et al., Phys. Plasmas **9**, 2214 (2002)
54. A.J. Mackinnon et al., Phys. Rev. Lett. **97**, 045001 (2006)
55. A. Ravasio et al., Phys. Rev. E **82**, 016407 (2010)
56. J.A. Cobble et al., J. App. Phys. **92**, 1775 (2002)
57. A.Y. Faenov et al., App. Phys. Lett. **95**, 101107 (2009)
58. C.I.I. Choi et al., J. Opt. Soc. Korea **13**, 28 (2009)
59. G. Sarri et al., New J. Phys. **12**, 045006 (2010)
60. A.J. Mackinnon et al., Rev. Sci. Instrum. **75**, 3531 (2004)
61. L. Romagnani et al., Phys. Rev. Lett. **95**, 195001 (2005)
62. K. Quinn et al., Phys. Rev. Lett. **102**, 194801 (2009)
63. L. Lancia et al., Phys. Rev. Lett. **113**, 235001 (2014)
64. J.J. Santos et al., New J. Phys. **17**, 083051 (2017)
65. W. Fox et al., Phys. Rev. Lett. **111**, 225002 (2013)
66. H. Ahmed et al., Phys. Rev. Lett. **110**, 205001 (2013)
67. H. Ahmed et al., ApJ Lett. **834**, L21 (2017)
68. M. Koenig et al., Plasma Phys. Control. Fusion **47**, B441 (2005)
69. A. Mancic et al., High Energy Density Phys. **6**, 21 (2010)
70. A. Pelka et al., Phys. Rev. Lett. **105**, 265701 (2010)
71. A. McKelvey et al., Sci. Rep. **7**, 7015 (2017)
72. B. Dromey et al., Nat. Comm. **7**, 10642 (2016)
73. L. Senje et al., App. Phys. Lett. **110**, 104012 (2017)
74. A. Yogo et al., App. Phys. Lett. **98**, 053701 (2011)
75. S.D. Kraft et al., New J. Phys. **12**, 085003 (2010)
76. D. Doria et al., AIP Adv. **2**, 011209 (2012)
77. J. Bin et al., App. Phys. Lett. **101**, 243701 (2012)
78. F. Hanton et al., Sci. Rep. **9**, 4471 (2019)
79. K. Zeil et al., App. Phys. B **110**, 437 (2013)
80. S. Rashke et al., Sci. Rep. **6**, 32441 (2016)
81. L. Manti et al., JINST **12**, C03084 (2017)
82. V. Favaudon et al., Sci. Transl. Med. **6**, 245ra93 (2014)
83. A. Yogo, Biological responses triggered by laser-driven ion beams, in *Laser-Driven Particle Acceleration Towards Radiobiology and Medicine* (Springer, 2016)
84. https://www.ptcog.ch/index.php/facilities-in-operation
85. W.D. Newhauser, R. Zhang, Phys. Med. Biol. **60**, R155 (2015)
86. D. Schardt, T. Elsasser, D. Schulz-Ertner, Rev. Mod. Phys. **82**, 383 (2010)
87. S.V. Bulanov, V. Khoroshkov, Plasma Phys. Rep. **28**, 453 (2002)
88. E. Fourkal et al., Med. Phys. **29**, 2788 (2002)
89. V. Malka et al., Med. Phys. **31**, 1587 (2004)
90. K.W.D. Ledingham, P.R. Bolton, N. Shikazono, C.-M. Ma, Appl. Sci. **4**, 402 (2014)
91. U. Masood et al., Phys. Med. Biol. **62**, 55331 (2017)
92. S. Schell, J.J. Wilkens, Med. Phys. **37**, 5330 (2010)
93. K.M. Hoffman, U. Masood, J. Pawelke, J.J. Wilkens, Med. Phys. **42**, 5120 (2015)
94. D. Margarone et al., Quantum Beam Sci. **2**, 8 (2018)
95. C. Obcemea (in this volume, 2018)
96. K.W.D. Ledingham, P. McKenna, R.P. Singhal, Science **300**, 1107 (2003)
97. F. Negoita et al., Rom. Rep. Phys. **68**, S37 (2016)
98. J.M. Yang et al., J. App. Phys. **96**, 6912 (2004)

99. L. Willingale et al., Phys. Plasmas **18**, 083106 (2011)
100. C. Zulick et al., App. Phys. Lett. **102**, 124101 (2013)
101. M. Roth et al., Phys. Rev. Lett. **110**, 044802 (2013)
102. S. Kar et al., New J. Phys. **18**, 053002 (2016)
103. M. Roth, Neutron generation, in *Applications of Laser-Driven Particle Acceleration* (CRC Press, 2018)
104. S.R. Mirfayzi et al., App. Phys. Lett. **111**, 044101 (2017)
105. M. Roth et al., Phys. Rev. Lett. **86**, 436 (2001)
106. M. Tabak et al., Phys. Plasmas **1**, 1626 (1994)
107. M. Temporal, J.J. Honrubia, S. Atzeni, Phys. Plasmas **9**, 3098 (2002)
108. J.J. Honrubia, M. Murakami, Phys. Plasmas **22**, 012703 (2015)
109. J.C. Fernandez et al., Nucl. Fus. **54**, 054006 (2014)
110. P. Mckenna et al., Phil. Trans. R. Soc. A, **364**, 711 (2006)
111. F. Wagner et al., Phys. Rev. Lett. **116**, 205002 (2016)

Chapter 8
Ultrafast Plasma Imaging

Malte C. Kaluza

Abstract This paper gives an overview of high-resolution diagnostic techniques, which can be used for ultrafast plasma imaging. Various effects in the plasma are exploited to realize diagnostics sensitive to density distributions (via interferometry) or small-scale internal plasma structures (shadowgraphy). Furthermore, magnetic field distributions, which are linked to the formation of a relativistic particle pulse, can be detected using polarimetry. After a short description of these effects possible experimental configurations are discussed and exemplary experimental results are presented, which highlight the great potential of such diagnostics for giving us high-resolution insights into laser-based particle accelerators.

8.1 Introduction

The acceleration of electrons and ions from relativistic plasmas generated by high-intensity laser pulses has attracted considerable attention over the last decades [13, 16, 22, 24]. Due to significant advances in high-power laser technology over the last couple of years, multi-100 Terawatt (TW) and Petawatt (PW) laser systems are now available in a growing number of laboratories all over the world, clearly in national research institutions but more and more also at universities. Focusing pulses from such laser systems on different types of targets ranging from low-density gas jets [30] or gas cells over liquid [35] and cryogenically cooled targets [9, 14] to solid foil or bulk targets generates transient plasmas in which electric field distributions with peak amplitudes of several 100 GV/m to 1 TV/m and more can be generated [32, 34] and further used to accelerate charged particles. The growing interest in particle accelerators driven by such high-power laser systems is also due to the fact that some of the parameters of the generated particle pulses are not only compatible but sometimes even superior to the parameters of particle pulses generated from large-scale conventional accelerators. Here, particle pulse parameters such as the pulse duration

M. C. Kaluza (✉)
Institute of Optics and Quantum Electronics, Max-Wien-Platz 1, 07743 Jena, Germany
e-mail: Malte.Kaluza@uni-jena.de

Helmholtz-Institute Jena, Fröbelstieg 3, 07743 Jena, Germany

© Springer Nature Switzerland AG 2019
L. A. Gizzi et al. (eds.), *Laser-Driven Sources of High Energy Particles and Radiation*,
Springer Proceedings in Physics 231, https://doi.org/10.1007/978-3-030-25850-4_8

(electron pulses can have durations as short as a few femtoseconds [6, 11, 21]) or the emittance or brilliance of the pulses (electron pulses: [5] ion pulses: [10]) are not or only under extreme precautions achievable with conventional particle accelerators. Another advantage of laser-driven particle accelerators comes from the availability of the afore mentioned electric field amplitudes in the plasma, reaching values up to 1 TV/m. Such field strengths, which cannot be generated in conventional accelerators due to ionization-induced material break-down in the accelerator structure, bear the potential of significantly reducing the acceleration length and hence the physical size of the accelerator. In fact, the current record of 7.8 GeV electron energies was achieved by accelerating the electrons over a distance of 20 cm only [15]. It is due to these advantages of laser-driven particle accelerators that they are envisaged as a potential future alternative to conventional particle accelerators with a number of different applications such as secondary radiation sources [27], material radiography, probes for ultra-fast, transient phenomena [4], materials research, inertial confinement fusion [26], medical applications [23] or as frontends for conventional accelerators [7, 8]

There are, however, a few challenges that currently still prevent such laser-driven plasma accelerators from becoming ready-to-use particle sources with broad applicability, e.g. in industry or medicine. Among these are the still limited peak energy, which becomes crucial e.g. when discussing the application of laser-driven ion beams in radiation therapy. Here, proton energies of 200–250 MeV would be necessary to reach and destroy deep-sited tumors located inside the human body. Even more severe, however, is the lack of controllability and shot-to-shot stability of the particle pulses, which are generated from subsequent, newly formed laser-generated plasmas. Since the parameters of the particle pulses strongly depend on the conditions of the plasma as the acceleration medium, a high degree of control over the parameters of these plasmas appears as a natural precondition to produce reproducible and stable particle bunches. Before having the chance to control this plasma, however, the diagnosis of the relevant plasma parameters, which are dominating the particle pulses' parameters appears as a natural first step. Keeping in mind that the laser-generated plasma rapidly evolves on length- and time-scales determined by the driving laser pulse, i.e. on few-fs time and few-μm length scales, it seems obvious that the diagnostics of the plasma need to have the potential to resolve the physical quantities and plasma parameters on these length and time scales too.

While there is a plethora of plasma diagnostics which can measure virtually all the relevant parameters of the plasma, such as temperature, density, electric- or magnetic-field distributions in a quasi-static manner, therefore mostly integrating over the whole interaction time, the availability of diagnostic methods having the potential of few-fs resolution is still rather limited [12]. Here, diagnostic techniques based on ultra-short electromagnetic pulses as so-called *probe pulses* have the potential to resolve also the fastest plasma dynamics. Usually, these probe pulses are split off of the main, driving laser pulse (also called the *pump pulse*), which leads to an intrinsic, almost perfect synchronization between pump and probe pulses. However, it is not straight-forward to detect all relevant plasma parameters, such as temperature or electric or magnetic field distributions using such pulses, since the sensitivity of the probe pulses to the different plasma parameters often also depends on the probe's wave length.

In this paper, we will review ultra-fast plasma imaging techniques based on the application of ultra-short probe pulses and experimental results from the application of such pulses in various scenarios for particle acceleration. While in Sect. 8.2, we will describe basic considerations and introduce effects which are underlying these diagnostics, Sect. 8.3 will describe possibilities to generate an optical probe pulse, which is necessary for realizing such diagnostics. Section 8.4 will then concentrate on possible implementations of such diagnostics and Sect. 8.5 will present a few exemplary experiments in which these diagnostics have been used in particle acceleration scenarios.

8.2 Physical Effects Relevant for Ultrafast Plasma Imaging

As discussed in the introduction, laser-driven plasma accelerators have intrinsic length and time scales on the μm- and fs-level, respectively. For a diagnostic, which needs to be able to resolve both scales at the same time, the application of ultra-short electromagnetic probe pulses seems a natural choice. Despite the spatial and temporal resolutions achievable with such pulses, in particular in combination with high-resolution imaging systems (which—in the case of visible or mid-infrared probe pulses—can reach a spatial resolution close to the wavelength of the applied probe radiation), it still needs to be discussed, which parameters and properties of the plasma can indeed be investigated when using electromagnetic radiation.

8.2.1 The Plasma's Refractive Index

First, let us consider an electromagnetic probe wave (with angular frequency ω_{pr} and wave vector $\mathbf{k}_{pr}$), which is propagating through a plasma of density n_e or—if the plasma density's spatial and temporal variation need to be taken into account—$n_e(\mathbf{r}, t)$. Here, we have assumed that the plasma's natural frequency, the so-called plasma frequency

$$\omega_{pl} = \sqrt{\frac{n_e e^2}{\varepsilon_0 m_e}}, \tag{8.1}$$

is smaller than the probe's angular frequency, i.e. $\omega_{pl} < \omega_{pr}$, i.e. the plasma is called underdense. In an overdense plasma with $\omega_{pl} > \omega_{pr}$ the probe pulse would not be able to propagate.[1] Due to the oscillation of the plasma electrons, which were induced

---

[1] Note that the local plasma frequency may become space and time dependent through the respective variation of n_e. Furthermore, once the plasma electrons start to move with relativistic speeds, e.g.

by the oscillating electric field of the probe pulse in the first place, these plasma electrons reemit radiation at the same frequency (like a forced oscillator). Depending on the ratio of the frequencies between the external oscillation (in our case the probe frequency ω_{pr}) and the resonance frequency of the oscillator (in our case the plasma frequency ω_{pl}), a phase shift between initial and induced oscillation occurs, which leads to a resulting wave superpositioned by both oscillations (which have the same frequencies) but with a slightly modified propagation speed of the oscillations, i.e. a slightly changed phase velocity. This phase velocity can be described by the refractive index η of the plasma with

$$\eta = \sqrt{1 - \frac{\omega_{pl}^2}{\omega_{pr}^2}} = \sqrt{1 - \frac{n_e}{n_{cr}}} < 1. \tag{8.2}$$

For the second expression, we have introduced the critical plasma density $n_{cr} = \omega_{pr}^2 \varepsilon_0 m_e / e^2$ for the probe pulse's wave length or frequency, i.e. the plasma density up to which the probe pulse can still propagate in the plasma before the latter becomes overdense for the probe light. With the plasma's refractive index the phase velocity is given by $v_\Phi = c/\eta > c$. In addition, the probe pulse's group velocity can be written as $v_{gr} = c \cdot \eta < c$. This relation between plasma density and probe pulse's phase velocity can be exploited, when the phase Φ accumulated by the probe wave propagating along a path with length L in x−direction through a plasma of density $n_e(x)$ is compared to the case that the same wave would have propagated through vacuum (where the refractive index $\eta_{vac} = 1$). The difference between these two phases, $\Delta\Phi$, is then given by

$$\Delta\Phi = \frac{2\pi}{k_{pr}} \int_L [\eta_{vac} - \eta(x)]\,dx = \frac{2\pi}{k_{pr}} \int_L \left[1 - \sqrt{1 - n_e(x)/n_{cr}}\right] dx \tag{8.3}$$

$$\approx \frac{\pi}{k_{pr} n_{cr}} \int_L n_e(x)dx. \tag{8.4}$$

Note that the last approximation is valid for a strongly underdense plasma, i.e. for $n_e(x)/n_{cr} \ll 1$. This phase difference between a probe wave propagating through a plasma and a wave going through vacuum can be measured using an interferometer, as it will be discussed in the next section.

Exploiting this effect, we are able to measure the electron density distribution in the plasma. Note, however, that the phase shift $\Delta\Phi$ is generated by a line integration, i.e. the probe wave accumulates the measureable phase shift along its entire path through the plasma. When one is interested in the spatially resolved electron density distribution (i.e. $n_e(\mathbf{r})$), certain assumptions about the symmetry of the plasma have to

---

through the interaction with a high-intensity laser pulse, relativistic corrections to the electron mass need to be taken into account, too.

be made that the density can be derived using an Abel inversion. If this is not possible (or if the degree of symmetry is unknown or needs to be checked) holographic methods have to be employed [1, 20].

8.2.2   Effects Sensitive to Magnetic Fields

Furthermore, since particle accelerators rely on electric and magnetic fields which accelerate the charged particles in the first place, the ability to measure electric and/or magnetic fields is also an important aspect. However, an electromagnetic probe pulse (as introduced above) propagating in vacuum will not directly be sensitive to electric or magnetic field distributions $\mathcal{B}(\mathbf{r}, t)$. This is in contrast to pulses of charged particles (e.g. electron or proton pulses moving with a velocity $\mathbf{v}$), which are deflected due to the Lorentz-force $\mathbf{F}(\mathbf{r}, t) = e \cdot (\mathcal{E} + \mathbf{v} \times \mathcal{B})$ acting on a single proton or electron of charge $q_p = +e$ and $q_e = -e$, respectively, in this beam. However, electro-magnetic pulses are susceptible, e.g. to magnetic field distributions, if these fields are present in a plasma of electron density $n_e(\mathbf{r}, t)$.[2] In this case, the Faraday effect or the Cotton-Mouton effect may play a role, depending on the orientation of the magnetic field lines and the direction (and polarization) of the probe pulse. For the case of a parallel orientation of the magnetic fields and the probe pulse's propagation direction (which is parallel to $\mathbf{k}_{pr}$), the response of the plasma electrons to the oscillating probe fields changes when compared to the case without magnetic fields. Since their forced oscillation leads to a motion perpendicular to the magnetic field, the induced Lorentz force alters their oscillation motion and—as a consequence—also the emitted radiation. In this case it is more intuitive to describe the initially linearly polarized probe light as the superposition of two circularly polarized waves. Here, one of them has the same sense of rotation as the electrons which—in the case they were free—would be carrying out cyclotron revolutions with the Larmor frequency $\omega_L = eB/m_e$. Due to this break of symmetry it is obvious that the two circularly polarized components are affected differently by the magnetized plasma. In fact, one of them moves with a slightly higher phase velocity than the other, which then leads to a continuous increase of the delay between the two circularly polarized components, when the light keeps propagating through the magnetized plasma. This increase in delay, however, is nothing but a continuous rotation of the plane of polarization of the probe beam, once we look at the linear oscillation again. This effect causing the polarization rotation is called the Faraday effect, and the rotation angle ϕ_{rot} in the plasma can be calculated by

$$\phi_{rot} = \frac{e}{2m_e c} \int_L \frac{n_e(\mathbf{r})}{n_{cr}} \mathcal{B}(\mathbf{r}) \cdot \frac{\mathbf{k}_{pr}}{k_{pr}} ds, \qquad (8.5)$$

---

[2]Note that the Faraday-effect also occurs in transparent media, which have a non-vanishing Verdet-constant.

where the integration again has to be carried out along the whole path of length L of the probe pulse through the plasma [33]. Once one is able to detect this change of the plane of polarization of the probe light, one is sensitive to detecting signatures of a magnetic field in the plasma. When the plasma density is measured simultaneously (e.g. by using interferometry), one can also deduce the spatial distribution of the magnetic field strength.

8.3 Generation of Synchronized Electromagnetic Probe Pulses

As it has been discussed in the previous sections, it is of great importance for laser-driven, plasma-based particle accelerators to have high-resolution diagnostics available for probing the evolution of the plasma and for measuring its parameters in a spatially and temporally resolved manner. For the latter point, it is of further importance to ensure the possibility to probe the status of the plasma at a fixed time during its evolution. Here, the use of probe pulses temporally synchronized to the driver pulse is essential. This can be ensured when the probe pulse is generated out of the main pulse, e.g. by using a partially transmissive mirror in the beam line [17] (cf. Fig. 8.1) or by using a pick-up mirror or a hole in a beam-line mirror.[3] If it is desired that the diagnostic has a temporal resolution better than the driver pulse's time scale (e.g. to be able to resolve processes occurring already during the driver pulses duration), it is not sufficient to simply use a split-off fraction of the main pulse (which would then lead to a time resolution similar to the driver pulse). Here, it is necessary to reduce the pulse duration, which can be accomplished by first broadening the probe pulse's spectrum e.g. via self-phase modulation in a gas-filled hollow core fibre and then recompressing this pulse, e.g. using chirped mirrors. This principle, as it is e.g. used at the JETI laser facility at IOQ and HI-Jena is shown in Fig. 8.1. In this example, the probe pulse can have a duration as short as (5.9 ± 0.4) fs [29] or—after optimizing the spectral broadening of 2.8 fs [2], which is significantly shorter than the driver pulse's duration of 32 fs. Using such a probe pulse allows e.g. for taking shadow-graphic snapshots of the plasma wave's evolution in a laser wakefield accelerator [28] where the probe's pulse duration below the driver pulse duration is essential. Furthermore, it is also possible to select a certain part of the ultra-broad spectrum of the probe pulse, which is different from the driver pulse spectrum. Using such a frequency shifted probe pulse light in combination with a well-adapted spectral filter (which effectively blocks out all other frequency components) light scattered

---

[3]Note that both approaches have their pros and cons: Using a partially transmissive beam-line mirror does not affect the driver beam profile, but the passage of the light through the mirror introduces spectral dispersion, which needs to be compensated. This becomes more and more challenging the broader the driver pulse's spectrum is. This can be avoided by using a pick-up mirror in the beam line (or by using a beam-line mirror with a small hole in the center). While this leads to an undisturbed probe pulse, it, however, likely leads to diffraction effects, which may affect the driver beam's profile.

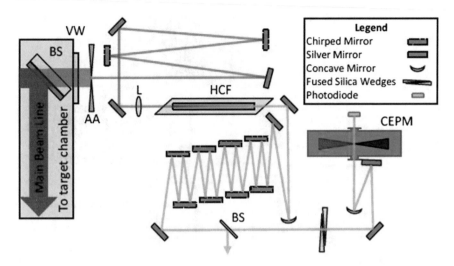

Fig. 8.1 Exemplary setup to generate few-cycle probe pulses synchronized to the main driver pulse [2]. Here, one of the mirrors in the driver pulse's beam line (BS) is partially transmissive through which a fraction of the beam is coupled out passing through a vacuum window (VW). This fraction is then first recompressed to its shortest duration (still similar to the driver pulse's duration) using chirped mirrors before it is focused by a lens (L) into a gas-filled hollow core fiber (HCF), where its spectrum is significantly broadened via self-phase modulation. After that, the pulses are compressed to their new Fourier-limited pulse duration using more reflexions off chirped mirrors. The duration of the probe pulses can be measured e.g. with a carrier-envelope phase meter (CEPM, [2]) or with an autocorrelator, before it is sent towards the target interaction chamber where it can diagnose the plasma generated by the driver pulse

from the interaction region (dominantly at the driver pulse's frequencies and harmonics) can be suppressed efficiently. Such scattered light could otherwise be much brighter than the probe light which would then outshine the probe light rendering time-resolved probe imaging impossible.

8.4 Specific Setups for Ultra-Fast Plasma Diagnostics

8.4.1 Interferometry

As it was discussed in Sect. 8.2.1, the plasma density is related to the plasma's refractive index η. Differences of the refractive index (e.g. when compared to vacuum) can be measured with a technique called interferometry. Such an interferometer can e.g. be set up in a Mach-Zehnder type geometry, as it is shown in Fig. 8.2. Here, a probe beam is split into two replica using a first beam splitter. One of the replica is propagating through the plasma (this is the signal arm), while the other part is covering the same geometrical distance but along another path which is not going through the

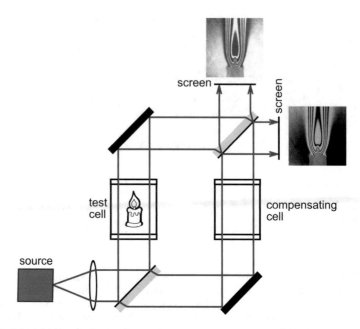

Fig. 8.2 Principle layout of a Mach-Zehnder type of interferometer. This image is reproduced from https://www.wikipedia.com

plasma (this is the reference arm). The two pulses are then recombined using a second beam splitter. The plasma can e.g. be imaged onto a CCD chip using a lens (or a more complex imaging system), which is positioned behind the second beam splitter. Then the two pulses will overlap and interfere on the camera. If no plasma is present and the two beams are recombined under a small tilt angle, parallel interference fringes will become visible in the image of the plasma. If, however, a plasma disturbs the signal beam the additional phase experienced by some parts of the probe wave will lead to a lateral bending of the interference fringes in the corresponding regions in the image. Note that in this scheme it is crucial that the signal and the reference arms have the same geometrical length—when neglecting the influence of the plasma—in order to allow for the two recombined replica to interfere on the CCD. This becomes all the more challenging the shorter the probe pulses are, since their longitudinal coherence length is determined (and hence limited) by their pulse duration. When using optical pulses with few-cycle duration for probing, the interference length is of the order of a few μm only, and the length of the two interferometer arms has to be aligned with an accuracy better than this.

Another, sometimes more elegant approach is to use a Nomarski-type interferometer, which employs a Wollaston prism [3]. A principle sketch is shown in Fig. 8.3. Here, the probe pulse does not need to be separated into two replica that have to be recombined after the interaction using beam splitters, but the pulse is split into two replica using a combination of two birefringent prisms, which are combined

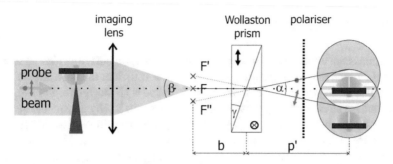

Fig. 8.3 Principle layout of a Nomarski-type interferometer using a Wollaston prism

in a so-called Wollaston prism. The two optical axes of the two prisms are aligned perpendicular with respect to each other and also perpendicular to the probe's direction of propagation. If the probe's linear polarization is initially aligned under an angle of 45° with respect to both optical axes, the probe pulse can be regarded as a combination of two linearly polarized parts, in Fig. 8.3 parallel and perpendicular to the plane of the drawing. When first entering the Wollaston prism, neither beam is diffracted. However, the polarization of one of the parts is parallel, that of the other one perpendicular to the optical axis in the first birefringent prism, i.e. one can be described as an ordinary ray, while the other one is an extraordinary ray. When entering the second prism, the ordinary ray becomes the extraordinary ray and vice versa. Therefore, both beams (with different polarizations) are refracted differently at this boundary, since one is passing from one medium with a smaller refractive index to one with a higher one and vice versa. When exiting the rear surface of the second prism, the two rays propagate in slightly different directions, which enclose the angle α, which is specific for a specific Wollaston prism. Assuming that a plasma is imaged onto a CCD using an imaging system, which we again simplify by a single imaging lens, then the initially collimated probe beam will be focused. Due to the refraction of the rays in the Wollaston prism the focal spot F appears to be separated into two virtual focal spots F' and F'', which are slightly separated in the lateral direction. Note that the two rays have different, in particular perpendicular polarizations. Once the angle β, which is related to the focusing of the collimated probe beam by the imaging lens, is larger than the separation angle α, the two diverging beams will always partially overlap. However, they will not interfere, since their polarizations are still perpendicular. If an additional polarizer is placed in the path of the two diverging beams and rotated by 45° with respect to both polarizations, the two beams will afterwards have the same polarization and intensity and can therefore interfere in the overlapping region. If the interaction region is placed in the one half of the probe beam, while the other half is propagating through undisturbed regions (i.e. vacuum), in the overlapping part of the two replica behind the Wollaston prism the disturbed part is interfering with the undisturbed part, i.e. we obtain an interferogram of the interaction region, which can then be analyzed to deduce the plasma density distribution.

8.4.2 Shadowgraphy

When one is not explicitly interested in the plasma density but in the plasma size and internal structures (e.g. a plasma wave), it is often sufficient to take a simple image of the plasma region only. Such an image is called a shadowgram, the associated technique is called shadowgraphy. Here, we just want to briefly discuss the question, why a plasma, which is only a phase object, can be seen in such an image. In an underdense plasma, the absorption of the probe light can to first order be neglected, i.e. it is not responsible for the formation of the image. However, the refractive index of the plasma and in particular its spatial variation leads to differences of the optical path length of different rays of the probe beam. Furthermore, if the refractive index of the plasma varies on small spatial scales, the rays of the probe beam can also experience diffraction. Therefore, the initially parallel probe rays will have been deflected by the plasma distribution. However, when imaging this plasma with a high-resolution objective, all rays which have been deflected should be collected and re-focused to the image plane, i.e. one would not necessarily expect that the image would show signatures of the plasma. However, since the depth of focus of the imaging optic is usually quite small (in particular when a high spatial resolution is desired), the probe rays will be deflected already before and also after the object plane. As a consequence, the different rays will cause interference in the image formed on the CCD chip (since the sources of the probe rays' deflection are no longer in the object plane). Therefore, even using a simple imaging system will allow us to take images of the plasma with a clear intensity variations caused by the plasma, which reflect size and—to some extent—the inner structure of the plasma. However, a quantitative analysis of the observed features may become rather complex, since there are no simple back-tracing algorithms from a certain image to a 3-dimensional plasma structure. However, when using numerical simulations (e.g. 3-dimensional particle-in-cell codes) it becomes possible to accurately model the formation of the image of the interaction which then helps to deduce some quantitative information about the plasma density distribution [19, 31].

8.4.3 Polarimetry

When one wants to employ the Faraday effect in a plasma in order to gain information about magnetic field distributions (which—as we have seen above—may lead to modifications of the probe pulse's polarization for the correct orientation of magnetic field lines and the probe pulse's propagation direction), one needs to modify the imaging setup (which has, e.g. been used to obtain shadowgrams before) in order to be sensitive to small polarization changes. Since a charge-coupled device (CCD) camera is usually sensitive to intensity variations but not to polarization changes, one can in principle simply place a polarizer in front of the CCD camera to translate the changes of the polarization into changes of the intensity on the CCD chip.

Let us first assume that the probe pulse is linearly polarized (which can in practice be realized by adding a high-quality polarized into the probe beam's path before it traverses the interaction) and that there is initially no interaction. If the probe pulse's intensity at a certain position (x_0, y_0) in the image plane is $I_0(x_0, y_0)$ and the polarizer is rotated away from maximum extinction—i.e. from the direction perpendicular to the orientation of the probe pulse's $\mathcal{E}$-field vector—by an angle of θ_{poll} then the probe's intensity $I_{poll}(x_0, y_0)$ transmitted through this polarizer is determined by Malus's law to

$$I_{poll}(x_0, y_0) = I_0(x_0, y_0)\left[1 - \beta_1 \sin^2(90° - \theta_{poll})\right], \quad (8.6)$$

where β_1 is the extinction ratio of this polarizer. If the probe's plane of polarization is furthermore rotated in the plasma due to the Faraday effect by an angle ϕ_{rot}, then this additional angle needs to be included in this expression yielding

$$I_{poll}(x_0, y_0) = I_0(x_0, y_0)\left[1 - \beta_1 \sin^2(90° - \theta_{poll} - \phi_{rot})\right]. \quad (8.7)$$

By doing so, a shadowgraphic image (or a shadowgram) becomes a polarogram. If the (initial) intensity distribution of the probe pulse in the plane, which is imaged by the imaging system onto the CCD chip, is known, taking a single polarogram would in principle be sufficient to extract the information about the polarization rotation.

In practice, however, the probe beam has a non-uniform intensity distribution, which is further modified by refraction and diffraction of the probe pulse when it traverses the plasma. Furthermore, since the imaging system produces a sharp image of one image plane only, the regions in front of this plane and behind it also contain plasma, which alters the probe pulse significantly due to interference. As a result, a simple shadowgram (i.e. taken without a polarizer) may already show a large variation of bright and dark regions. Furthermore, since the plasma is often subject to non-linear evolutions, the modifications to the probe beam may significantly change from shot to shot.

To eliminate intensity variations caused by the initial (in general non-uniform) near-field intensity profile of the probe and of diffraction and refraction of the probe rays in the plasma, two images of the same interaction region can be taken simultaneously using two different CCD cameras, each equipped with its own polarizer (then set to the two angles θ_{poll} and θ_{pol2}) separated by a non-polarizing beam splitter. Such a setup is schematically shown in Fig. 8.4. Here, intensity variations induced by refraction and diffraction of the probe rays in the plasma appear in both images and can be eliminated by a direct comparison of these two images. Furthermore, this method does not suffer from shot-to-shot variations, since the two images can be recorded on the same shot.

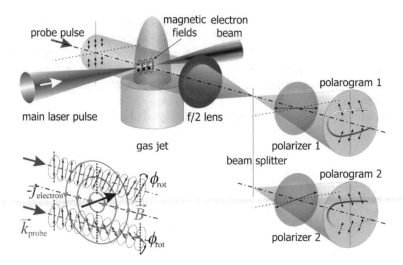

Fig. 8.4 Sketch of an experimental setup to probe magnetic field distributions in a laser-generated plasma using an initially linearly polarized probe pulse [18]. The probe pulse's polarization is modified by magnetic field distributions in the plasma via the Faraday effect. Imaging the same region in the plasma on two CCD cameras by using a non-polarizing beam splitter but by equipping each CCD camera with an individual polarizer (set to two different angles θ_{pol1} and θ_{pol2}) makes this technique sensitive to changes in the polarization (which are different in the two polarograms) but insensitive to intensity variations in the probe pulse due to diffraction or refraction (which are identical in the two polarograms). The inset in the lower left corner shows the principle of the Faraday effect which is responsible for the polarization rotation in the plasma

8.5 Experimental Examples for Ultra-Fast Plasma Probing

In this section, we want to present a few examples, where the diagnostic techniques that were introduced in the previous sections, have been used to gain deeper insight into the interaction between the laser and the plasma. Please keep in mind that this selection is by far not exhaustive. For a comprehensive review on diagnostic techniques and their application, we refer e.g. to the review paper by Downer et al. [12].

8.5.1 Measurement of the Plasma Density Using Interferometry

Here, we show exemplary images from the interaction of two separate laser beams with a thin plastic foil which was then transversely probed by a synchronized, frequency doubled probe pulse. The two laser pulses initiating the interaction were (i) a few-ns long, frequency-doubled laser pulse from a Nd:glass laser system delivering

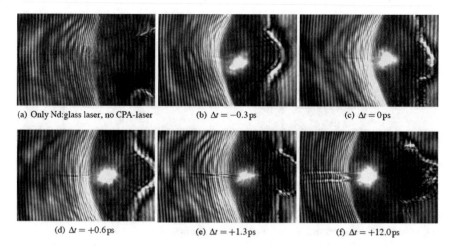

(a) Only Nd:glass laser, no CPA-laser   (b) $\Delta t = -0.3\,$ps   (c) $\Delta t = 0\,$ps

(d) $\Delta t = +0.6\,$ps   (e) $\Delta t = +1.3\,$ps   (f) $\Delta t = +12.0\,$ps

Fig. 8.5 Series of interferograms to measure the plasma density distribution of the interaction of a high-intensity CPA main pulse (coming from the left) with a preplasma in front of a thin plastic foil generated by a frequency-doubled, ns-long laser pulse with 10 J energy (also coming from the left). While in panel (**a**), there was no high-power laser pulse, i.e. only the preplasma is measured, panels (**b**)–(**f**) show the interaction of the main pulse with this preplasma at different times during the evolution, realized by varying the relative delay between the CPA main pulse and the probe pulse

up to 10 J energy into a focal spot of 200 μm diameter which produced a preplasma on the front surface of the plastic foil with a long scale length. Into this preplasma, the second laser pulse (delivering up to 800 mJ of energy in a duration of 150 fs) was focused to a spot of a few μm in diameter. This second, high-power laser pulse generated a plasma channel in the preplasma, in which electrons were accelerated into the foil. Using a synchronized, frequency doubled probe pulse (also of 150 fs duration but at 395 nm wave length) time-resolved snap shots of the channel formation in the preplasma could be taken. By changing the delay between the high-intensity main pulse and the synchronized probe pulse, a time-sequence of images could be taken as it is shown in Fig. 8.5. Using a Wollaston prism interferometric images could be taken to deduce the electron density. While panel (a) shows the preplasma alone (i.e. without the high-intensity main pulse), the subsequent images show the formation and later lateral expansion of the plasma channel for different times. Note that the bright spot in the center of panels (b)–(f) is due to plasma emission at the second harmonic of the high-intensity main pulse.

While such images can already give a qualitative idea about the preplasma extent and—in this particular case—the evolution of the plasma channel in the preplasma induced by the interaction with the CPA main pulse, such interferograms can also be used to deduce the preplasma density distribution quantitatively. This is particularly easy to do in the case that the preplasma can be assumed to have cylindrical symmetry. This is usually a valid assumption if the preplasma was generated by a laser pulse irradiating a plane target under normal incidence—as it was the case for the ns-laser

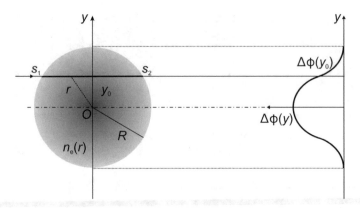

Fig. 8.6 Schematic sketch for the calculation of the (radially symmetric) plasma density distribution in a transverse probing geometry when applying an Abel inversion

pulse which generated the preplasma in the images presented here.[4] In this situation, the axis of symmetry is the target normal going through the center of the focal spot on the target front surface. One can use an Abel-transformation to deduce the plasma density distribution, as it is schematically shown in Fig. 8.6. Here, the gray shaded area schematically shows a cut through the preplasma in a plane parallel to the target surface. The axis of symmetry of this distribution goes through the origin ($\mathcal{O}$) of the coordinate system. A single ray of the probe beam (which propagates from left to right in this image) traverses the plasma distribution and acquires an optical phase as dictated by the plasma's refractive index. When compared to the case that a similar ray would have propagated through vacuum (along a path of equal geometrical length but without a plasma) a phase difference $\Delta\phi(y_0)$ is aquired, which depends on the lateral position y_0 of the ray. This phase difference is related to the local refractive index $\eta(x, y_0)$, which the ray experiences on its path through the plasma. When assuming that the plasma's refractive index $\eta = 1$ beyond the boundary with the radius R from the origin, (or—in other words—the ray propagates through vacuum beyond the radius R) the phase difference is

$$\Delta\phi(y_0) = \frac{2\pi}{\lambda_{\text{pr}}} \int_{x_1}^{x_2} [1 - \eta(x, y_0)]\mathrm{d}x, \qquad (8.8)$$

---

[4]One should note, however, that any deviations from this cylindrical symmetry cannot be resolved with the simplest approach of an Avel inversion, which inherently assumes this symmetry. Such asymmetries can—for example—be caused by a non-symmetric focus or—when using a gas jet for producing an underdense plasma—a non-symmetric gas nozzle. For various applications, a non-symmetric density distribution may even be advantageous. For more accurate results, which can resolve also non-symmetric density disrtibutions, tomographic techniques would be required [1, 20].

where x_1 and x_2 indicate the positions where the ray enters and exits the plasma region, respectively. For a low density plasma with

$$\eta = \sqrt{1 - n_e/n_{cr}} \approx 1 - n_e/2n_{cr}, \tag{8.9}$$

this expression—when used in (8.8)—leads to

$$\Delta\phi(y_0) \approx \frac{\pi}{n_{cr}\lambda_{pr}} \int_{x_1}^{x_2} n_e(x, y_0)dx = \frac{2\pi}{n_{cr}\lambda_{pr}} \int_{y_0}^{R} \frac{n_e(r)r}{\sqrt{r^2 - y_0^2}}dr, \tag{8.10}$$

where we have substituted the integration variable according to $x = \sqrt{r^2 - y_0^2}$ and $dx = rdr/\sqrt{r^2 - y_0^2}$. Now the radially symmetric plasma density $n_e(r)$ can be deduced via an Abel inversion to

$$n_e(r) = -\frac{n_{cr}\lambda_{pr}}{\pi^2} \int_r^R \frac{d}{dy}\Delta\phi(y) \cdot \frac{dy}{\sqrt{y^2 - r^2}}. \tag{8.11}$$

When applying this method to an interferogram from the same sequence as shown above the density distribution shown in Fig. 8.7 can be calculated. Here, the density distribution from the area marked by the red boundary is shown in the right part of

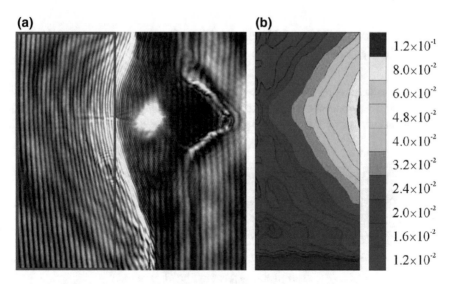

	1.2×10^{-1}
	8.0×10^{-2}
	6.0×10^{-2}
	4.8×10^{-2}
	4.0×10^{-2}
	3.2×10^{-2}
	2.4×10^{-2}
	2.0×10^{-2}
	1.6×10^{-2}
	1.2×10^{-2}

Fig. 8.7 Analysis of an interferogram. The region indicated by the red box in (**a**) is analyzed and yields an electron density distribution as shown in (**b**). The densities are given as fractions of the critical density for the probe wavelength of $\lambda_{pr} = 395$ nm

the image and given in units of the critical density of the probe wavelength (here, $\lambda_{pr} = 395$ nm, i.e. $n_{cr} = 7.2 \times 10^{21}/cm^3$).

8.5.2 High-Resolution Shadowgrams of the Plasma Wave in a Laser-Wakefield Accelerator

When a few-cycle probe pulse together with a high-resolution imaging system is used for probing the interaction of a high-power laser pulse with an underdense plasma, it may be possible to record the plasma wave, which is the central acceleration structure in a laser-wakefield accelerator [25, 34]. An exemplary image is shown in Fig. 8.8, where a plasma wave was driven by a pulse from the JETI laser at the IOQ in Jena, Germany in a plasma with a density of 1.5×10^{19} cm^{-3} [29]. In this image, the driving laser pulse is propagating from left to right and it has generated a periodic structure of density oscillations in the background plasma (the so-called plasma wave). The driving laser pulse is positioned at the longitudinal position of $0\,\mu$m. In the first part of the plasma wave (i.e. in the oscillations directly following the driving laser pulse), the shape of the individual plasma wave periods resembles a horse-shoe, which is indicative of relativistic effects and the reduction of the plasma density on the laser axis due to ponderomotive effects. Further behind, the plasma wave periods evolve into a triangular shape which is indicative of transverse wave breaking.

8.5.3 Measurements of Magnetic Field Structures in a Laser-Wakefield Accelerator

In an experiment using the JETI-laser at IOQ in Jena, Germany, in which a laser-wakefield accelerator was employed to produce quasi-monoenergetic electron pulses, the magnetic field structure inside the plasma was measured using a transverse optical

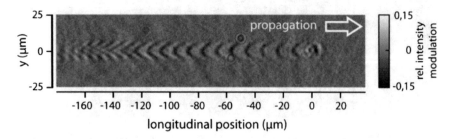

Fig. 8.8 Exemplary image of a laser-driven plasma wave, which was obtained by taking a shadowgraphic image using a few-cycle optical probe pulse and a high-resolution imaging system [29]

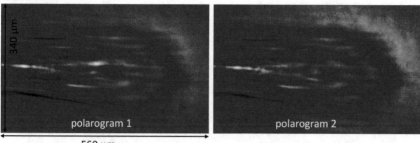

Fig. 8.9 Two polarograms (shown in false colour) taken during the same laser shot but with polarizers aligned differently (see text). While the strongly modulated structure is mostly due to diffraction and refraction of the probe pulse on its passage through the plasma the regions in the center of the images, where clear differences in colour (i.e. in intensity in the images) are visible

probe pulse (of 100 fs duration and at a wavelength of $\lambda_{pr} = 800$ nm) and employing the Faraday effect [18]. In a setup, similar to the one described in Sect. 8.4.3, two polarograms of the same interaction region in the plasma were taken on each laser shot. The two polarizers were rotated in opposite direction from the orientation of maximum extinction (here, $\theta_{pol1} = +5.9°$ and $\theta_{pol2} = -4.1°$). Two exemplary images of the interaction are shown in Fig. 8.9. As it was described in Sect. 8.2.2, the polarization of a probe pulse propagating through a magnetized plasma can be altered by the Faraday effect. As it is described by the scalar product between the local magnetic field $\mathcal{B}$ and the probe pulse's propagation direction linked to $\mathbf{k}_{pr}$ in 8.5, the rotation depends on the orientation of these two vectors. In particular, if one of them reverses its orientation, the sense of rotation of the probe's polarization will reverse too. In a situation, where the probe pulse propagates perpendicular to the direction of an electron current (as which an accelerated multi-MeV electron bunch can be described), which is associated with azimuthal magnetic fields, the probe will have an orientation parallel or anti-parallel to the $\mathcal{B}-$field lines depending on if it is passing the current below or above its axis. Therefore, the probe's polarization will be rotated either clockwise or counter-clockwise above and below the axis. If the two polarizers are also rotated in opposite directions, the sensitivity of the measurement will be improved. Regions of azimuthal magnetic fields will show up in the images as pairs of bright(er) and dark(er) patches, which should be symmetrical to the laser axis, which coincides with the propagation axis of the electron pulse (Fig. 8.10).

If the two polarograms shown in Fig. 8.9 are analyzed by carrying out a pixel-by-pixel division of the intensities, i.e. $I_{pol1}(x_0, y_0)/I_{pol2}(x_0, y_0)$, the intensity distribution $I_0(x_0, y_0)$ from (8.6), which contains all intensity variations of the probe excluding magnetic field effects, can be eliminated. From this intensity ratio the corresponding rotation angle ϕ_{rot} can be deduced numerically using the polarizer angles θ_{pol1} and θ_{pol2} from the experiment. The region in the plasma, where clear changes of the polarization angle are visible corresponds to regions of strong magnetic fields. A comparison with numerical simulations confirmed that these fields are

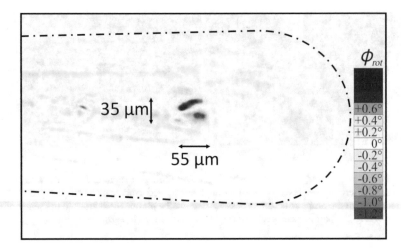

Fig. 8.10 The pixel-by-pixel intensity ratio of the two polarograms shown in Fig. 8.9 can be translated into the Faraday-rotation angle using the polarizer angles from the experiment. The region where clear rotation angles are visible corresponds to the magnetic field region caused by the MeV electron pulse together with the magnetic fields from the plasma wave

indeed caused by the azimuthal magnetic fields associated with the MeV electron pulse and the plasma wave. By changing the delay between pump and probe pulse in this experiment, the formation of the magnetic field feature and—since this was directly related to this—the injection and further acceleration of a relativistic electron pulse was possible. In this experiment, the region showing the strongest magnetic fields had a spatial extent of approximately $35\,\mu\mathrm{m} \times 55\,\mu\mathrm{m}$, which is much larger than one would expect. This, however, could be explained by the limited spatial and temporal resolution of the probing diagnostic available in this experiment. In fact, the spatial resolution was limited to about $10\,\mu\mathrm{m}$ in the transverse and $30\,\mu\mathrm{m}$ in the longitudinal direction. The latter value was also affected by the motion blur in the image caused by the motion of the main pulse (and the plasma wave) perpendicular to the propagation direction of the probe pulse.

8.6 Summary

This paper has given a short (and by far not exhaustive) overview of ultra-fast diagnostics which can be applied to laser-generated plasmas in the context of plasma-based particle accelerators. When employing the various diagnostic techniques, these measurements will be sensitive to various plasma quantities (e.g. density, magnetic fields, etc.). If synchronized few-cycle probe pulses are used in combination with high-resolution imaging setups, the plasma properties can be investigated in great detail. In the future, such diagnostic techniques bear the potential to give us more detailed insights into the physics underlying laser-based particle accelerators, potentially improving their performance, which is mandatory for the various envisioned applications of this novel type of particle accelerators.

References

1. A. Adelmann, B. Hermann, R. Ischebeck, M. Kaluza, U. Locans, N. Sauerwein, R. Tarkeshian, Real-time tomography of gas-jets with a Wollaston interferometer. Appl. Sci. **8**(3), 443 (2018)
2. D. Adolph, M. Möller, J. Bierbach, M.B. Schwab, A. Sävert, M. Yeung, A.M. Sayler, M. Zepf, M.C. Kaluza, G.G. Paulus, Real-time, single-shot, carrier-envelope-phase measurement of a multi-terawatt laser. Appl. Phys. Lett. **110**(8), 081105 (2017)
3. R. Benattar, C. Popovics, R. Sigel, Polarized light interferometer for laser fusion studies. Rev. Sci. Instrum. **50**(12), 1583–1586 (1979)
4. M. Borghesi, S. Bulanov, D.H. Campbell, R.J. Clarke, T.Z. Esirkepov, M. Galimberti, L.A. Gizzi, A.J. Mackinnon, N.M. Naumova, F. Pegoraro, H. Ruhl, A. Schiavi, O. Willi, Macroscopic evidence of soliton formation in multiterawatt laser-plasma interaction. Phys. Rev. Lett. **88**(13), 135002 (2002)
5. E. Brunetti, R.P. Shanks, G.G. Manahan, M.R. Islam, B. Ersfeld, M.P. Anania, S. Cipiccia, R.C. Issac, G. Raj, G. Vieux, G.H. Welsh, S.M. Wiggins, D.A. Jaroszynski, Low emittance, high brilliance relativistic electron beams from a laser-plasma accelerator. Phys. Rev. Lett. **105**(21), 215007 (2010)
6. A. Buck, M. Nicolai, K. Schmid, C.M.S. Sears, A. Sävert, J.M. Mikhailova, F. Krausz, M.C. Kaluza, L. Veisz, Real-time observation of laser-driven electron acceleration. Nat. Phys. **7**(7), 543–548 (2011)
7. S. Busold, A. Almomani, V. Bagnoud, W. Barth, S. Bedacht, A. Blažević, O. Boine-Frankenheim, C. Brabetz, T. Burris-Mog, T.E. Cowan, O. Deppert, M. Droba, H. Eickhoff, U. Eisenbarth, K. Harres, G. Hoffmeister, I. Hofmann, O. Jäckel, R. Jaeger, M. Joost, S.D. Kraft, F. Kroll, M.C. Kaluza, O. Kester, Z. Lecz, T. Merz, F. Nürnberg, H. Al-Omari, A. Orzhekhovskaya, G.G. Paulus, J. Polz, U. Ratzinger, M. Roth, G. Schaumann, P. Schmidt, U. Schramm, G. Schreiber, D. Schumacher, T. Stöhlker, A. Tauschwitz, W. Vinzenz, F. Wagner, S. Yaramyshev, B. Zielbauer, Shaping laser accelerated ions for future applications—the LIGHT collaboration. Nucl. Instrum. Methods Phys. Res. A **740**(C), 94–98 (2014)
8. S. Busold, D. Schumacher, C. Brabetz, D. Jahn, F. Kroll, O. Deppert, U. Schramm, T.E. Cowan, A. Blažević, V. Bagnoud, M. Roth, Towards highest peak intensities for ultra-short MeV-range ion bunches. Sci. Rep. 1–7 (2015)
9. R.A. Costa Fraga, A. Kalinin, M. Kühnel, D.C. Hochhaus, A. Schottelius, J. Polz, M.C. Kaluza, P. Neumayer, R.E. Grisenti, Compact cryogenic source of periodic hydrogen and argon droplet beams for relativistic laser-plasma generation. Rev. Sci. Instrum. **83**(2), 025102 (2012)
10. T.E. Cowan, J. Fuchs, H. Ruhl, A.J. Kemp, P. Audebert, M. Roth, R.B. Stephens, I. Barton, A. Blažević, E. Brambrink, J.A. Cobble, J.C. Fernández, J.C. Gauthier, M. Geissel, B.M. Hegelich, J. Kaae, S. Karsch, G.P. Le Sage, S. Letzring, M. Manclossi, S. Meyroneinc, A. Newkirk, H. Pépin, N. Renard-Le Galloudec, Ultralow emittance, multi-MeV proton beams from a laser virtual-cathode plasma accelerator. Phys. Rev. Lett. **92**(20), 204801 (2004)
11. A.D. Debus, M. Bussmann, U. Schramm, R. Sauerbrey, C.D. Murphy, Z. Major, R. Hörlein, L. Veisz, K. Schmid, J. Schreiber, K.J. Witte, S.P. Jamison, J.G. Gallacher, D.A. Jaroszynski, M.C. Kaluza, B. Hidding, S. Kiselev, R. Heathcote, P.S. Foster, D. Neely, E.J. Divall, C.J. Hooker, J.M. Smith, K. Ertel, A.J. Langley, P. Norreys, J.L. Collier, S. Karsch, Electron bunch length measurements from laser-accelerated electrons using single-shot THz time-domain interferometry. Phys. Rev. Lett. **104**(8), 084802 (2010)
12. M.C. Downer, R. Zgadzaj, A.D. Debus, U. Schramm, M.C. Kaluza, Diagnostics for plasma-based electron accelerators. Rev. Modern Phys. **90**(3), 035002 (2018)

13. E. Esarey, C.B. Schroeder, W.P. Leemans, Physics of laser-driven plasma-based electron accelerators. Rev. Modern Phys. **81**(3), 1229–1285 (2009)
14. M. Gauthier, J.B. Kim, C.B. Curry, B. Aurand, E.J. Gamboa, S. Göde, C. Goyon, A. Hazi, S. Kerr, A. Pak, A. Propp, B. Ramakrishna, J. Ruby, O. Willi, G.J. Williams, C. Rödel, S.H. Glenzer, High-intensity laser-accelerated ion beam produced from cryogenic micro-jet target. Rev. Sci. Instrum. **87**(11), 11D827 (2016)
15. A.J. Gonsalves, K. Nakamura, J. Daniels, C. Benedetti, C. Pieronek, T.C.H. de Raadt, S. Steinke, J.H. Bin, S.S. Bulanov, J. van Tilborg, C.G.R. Geddes, C.B. Schroeder, C. Toth, E. Esarey, K. Swanson, L. Fan-Chiang, G. Bagdasarov, N. Bobrova, V. Gasilov, G. Korn, P. Sasorov, W.P. Leemans, Petawatt laser guiding and electron beam acceleration to 8 GeV in a laser-heated capillary discharge waveguide. Phys. Rev. Lett. **122**(8), 084801 (2019)
16. S.M. Hooker, Developments in laser-driven plasma accelerators. Nat. Photonics **7**, 775 (2013)
17. M.C. Kaluza, M.I.K. Santala, J. Schreiber, G.D. Tsakiris, K.J. Witte, Time-sequence imaging of relativistic laser–plasma interactions using a novel two-color probe pulse. Appl. Phys. B **92**(4), 475–479 (2008)
18. M.C. Kaluza, H.P. Schlenvoigt, S.P.D. Mangles, A.G.R. Thomas, A.E. Dangor, H. Schwoerer, W.B. Mori, Z. Najmudin, K. Krushelnick, Measurement of magnetic-field structures in a laser-wakefield accelerator. Phys. Rev. Lett. **105**, 115002 (2010)
19. M.F. Kasim, L. Ceurvorst, N. Ratan, J. Sadler, N. Chen, A. Sävert, R.M.G.M. Trines, R. Bingham, P.N. Burrows, M.C. Kaluza, P. Norreys, Quantitative shadowgraphy and proton radiography for large intensity modulations. Phys. Rev. E **95**(2), 94–9 (2017)
20. B. Landgraf, M. Schnell, A. Sävert, M.C. Kaluza, C. Spielmann, High resolution 3D gas-jet characterization. Rev. Sci. Instrum. **82**(8), 083106 (2011)
21. O. Lundh, J.K. Lim, C. Rechatin, L. Ammoura, A. Ben-Ismaïl, X. Davoine, G. Gallot, J.P. Goddet, E. Lefebvre, V. Malka, J. Faure, Few femtosecond, few kiloampere electron bunch produced by a laser-plasma accelerator. Nat. Phys. **6**(12), 1–4 (2011)
22. A. Macchi, M. Borghesi, M. Passoni, Ion acceleration by superintense laser-plasma interaction. Rev. Modern Phys. **85**(2), 751–793 (2013)
23. V. Malka, S. Fritzler, E. Lefebvre, E. d'Humières, R. Ferrand, G. Grillon, C. Albaret, S. Meyroneinc, J.P. Chambaret, A. Antonetti, D. Hulin, Practicability of protontherapy using compact laser systems. Med. Phys. **31**(6), 1587–1592 (2004)
24. G.A. Mourou, T. Tajima, S.V. Bulanov, Optics in the relativistic regime. Rev. Modern Phys. **78**(2), 309–371 (2006)
25. A. Pukhov, J. Meyer-ter Vehn, Laser wake field acceleration: the highly non-linear broken-wave regime. Appl. Phys. B **74**, 355–361 (2002)
26. M. Roth, T.E. Cowan, M.H. Key, S.P. Hatchett, C.G. Brown, W. Fountain, J. Johnson, D.M. Pennington, R.A. Snavely, S.C. Wilks, K. Yasuike, H. Ruhl, F. Pegoraro, S.V. Bulanov, E.M. Campbell, M.D. Perry, H. Powell, Fast ignition by intense laser-accelerated proton beams. Phys. Rev. Lett. **86**(3), 436–439 (2001)
27. A. Rousse, K.T. Phuoc, R. Shah, A. Pukhov, E. Lefebvre, V. Malka, S. Kiselev, F. Burgy, J.P. Rousseau, D. Umstadter, D. Hulin, Production of a keV X-ray beam from synchrotron radiation in relativistic laser-plasma interaction. Phys. Rev. Lett. **93**(13), 135005 (2004)
28. A. Sävert, S.P.D. Mangles, M. Schnell, E. Siminos, J.M. Cole, M. Leier, M. Reuter, M.B. Schwab, M. Möller, K. Poder, O. Jäckel, G.G. Paulus, C. Spielmann, S. Skupin, Z. Najmudin, M.C. Kaluza, Direct observation of the injection dynamics of a laser Wakefield accelerator using few-femtosecond shadowgraphy. Phys. Rev. Lett. **115**(5), 055002 (2015)
29. M.B. Schwab, A. Sävert, O. Jäckel, J. Polz, M. Schnell, T. Rinck, L. Veisz, M. Möller, P. Hansinger, G.G. Paulus, M.C. Kaluza, Few-cycle optical probe-pulse for investigation of relativistic laser-plasma interactions. Appl. Phys. Lett. **103**(19), 191118 (2013)
30. S. Semushin, V. Malka, High density gas jet nozzle design for laser target production. Rev. Sci. Instrum. **72**(7), 2961–2965 (2001)
31. E. Siminos, S. Skupin, A. Sävert, J.M. Cole, S.P.D. Mangles, M.C. Kaluza, Modeling ultrafast shadow graphy in laser-plasma interaction experiments. Plasma Phys. Control. Fusion **58**, 065004 (2016)

32. R.A. Snavely, M.H. Key, S.P. Hatchett, T.E. Cowan, M. Roth, T.W. Phillips, M.A. Stoyer, E.A. Henry, T.C. Sangster, M.S. Singh et al., Intense high-energy proton beams from petawatt-laser irradiation of solids. Phys. Rev. Lett. **85**(14), 2945–2948 (2000)
33. J.A. Stamper, B.H. Ripin, Faraday-rotation measurements of megagauss magnetic fields in laser-produced plasmas. Phys. Rev. Lett. **34**(3), 138 (1975)
34. T. Tajima, J.M. Dawson, Laser electron accelerator. Phys. Rev. Lett. **43**, 267 (1979)
35. S. Ter-Avetisyan, M. Schnürer, S. Busch, E. Risse, P.V. Nickles, W. Sandner, Spectral dips in ion emission emerging from ultrashort laser-driven plasmas. Phys. Rev. Lett. **93**(15), 155006 (2004)

Chapter 9
Particles Simulation Through Matter in Medical Physics Using the Geant4 Toolkit: From Conventional to Laser-Driven Hadrontherapy

G. A. P. Cirrone, G. Cuttone, L. Pandola, D. Margarone and G. Petringa

Abstract Monte Carlo simulation represents nowadays one of the powerful approach for the simulation of very complex environments like those typical of medical physics where, in general, an accurate simulation of the involved radiation beams and of the patients are required to fully reproduce a clinical case. It since from 1963, when Berger introduced the condensed approach for the simulation of electron interaction with matter grew being today one of the most important tools used to verify the dose distribution in patients, design radiotherapy facility, study the radioisotopesproton/ion beams and their application (Berger. Monte Carlo calculation of the penetration and diffusion of fast charged particles, vol. I. Academic Press, New York, pp. 135–215, 1963 [1]),(Berger and Hubbell. XCOM: photon cross sections on a personal computer, Technical Report NBSIR 87–3597. National Institute of Standards and Technology, Gaithersburg, MD, 1987. [2]). Monte Carlo is also often used to evaluate important parameters related with the quality of a radiation treatment [3]. Evaluation of radiobiological damage from charged particles represents a complex calculation where a simple analytical approach is not sufficient for a precise and complete description of involved phenomena. In this work we will present, after a brief introduction on Monte Carlo method, the use of the open-source Geant4 toolkit for the simulation of a typical hadrontherapy passive beamline and how it can be efficiently used to retrieve critical parameters like LET and RBE.

G. A. P. Cirrone · G. Cuttone · L. Pandola (✉) · G. Petringa
INFN-LNS, Via S Sofia 62, 95123 Catania, Italy
e-mail: pandola@lns.infn.it

G. Petringa
e-mail: giada.petringa@lns.infn.it

G. A. P. Cirrone · D. Margarone
ELI-Beamline Project, Institute of Physics, ASCR, PALS Center, Prague, Czech Republic

© Springer Nature Switzerland AG 2019   187
L. A. Gizzi et al. (eds.), *Laser-Driven Sources of High Energy Particles and Radiation*,
Springer Proceedings in Physics 231, https://doi.org/10.1007/978-3-030-25850-4_9

9.1 Introduction

9.1.1 The Monte Carlo Approach

The Monte Carlo (MC) is a mathematical approach aiming to model nature through direct simulation of the essential dynamics of a given system. In the framework of physics processes studies, the Monte Carlo is used, for example, to model a macroscopic system through the simulation of its microscopic components.

Based on the random numbers generation and belonging to the so-called family of statistic and not-parametric methods, the term Monte Carlo was coined in 1947, by the scientists J. Von Neumann and S. Ulam, at the starting age of computers [4, 5]. They in fact, associated the intrinsic stochastic nature of the approach with the Monte Carlo city, being a roulette an almost perfect analogic random number generator.

In order to find a solution using the Monte Carlo three fundamental steps should be accomplished [6, 7]:

1. The generation of a sequence a random numbers and the determination of the input variable, depending on the chosen probability distribution;
2. The calculation of the output parameter;
3. The loop on previous points and comparison of the results in order to establish the minimum value of the associated variance.

The randomness of the generated numbers is a key requirement for the reliability of any Monte Carlo result. The number sequence generated by a calculator is defined to be pseudo-random because a pre-established relation between a sampled number and the previous one exists [3]. The entire sequence depends on the first number of the chain which is called seed.

A huge list of simulation codes, based on a Monte Carlo approach, have been developed and adopted in many application fields. Many of these have been specifically developed and publicly distributed to permit the simulation of particles interaction and propagation inside matter. Table 9.1 shows some of the most widespread Monte Carlo codes, the particles they are able to handle and the programming language used for their development [8–10].

9.2 Monte Carlo Simulation in Medical Physics

The history of the Monte Carlo method applied to the medical physics field is tied to the development of methods for the simulation of electron transport in complex geometries and in the description of electromagnetic cascades. In the 1950, Robert Wilson published the first papers describing the Monte Carlo method in the electron transport [11]. The growth of the use of linear electron accelerators (LINACs) for radiotherapy also pushed-up the development of Monte Carlo methods for the dose calculation. Since that time, the deployment in Monte Carlo radiation transport

Table 9.1 List of the most widespread Monte Carlo codes used for particle tracking

Code name	Particles handled	Language
ETRAN/ITS	Protons, Electrons	Fortran
PHITS	All particles	Fortran
SHIELD-HIT	All particles	Fortran
MCNP	Protons, Electrons, Neutrons	Fortran
MCNPX	All particles	Fortran
EGS	Photons, Electrons	Fortran
PENELOPE	Photons, Electrons	Fortran
GEANT3	All particles	Fortran
GEANT4	All particles	C++
FLUKA	All particles	Fortran
PETRA	Protons, Electrons, Neutrons	C++

algorithms has had an important impact in different areas of radiation dosimetry. For the experimental determination of absorbed dose, for example, several quantities are difficult to estimate accurately without numerical models. Radiation dosimetry detectors are in fact mostly constituted by several parts and different materials. In this context, Monte Carlo simulations can solve the problem related to the change in energy loss inside the materials [12]. Over the recent generations of radiation dosimetry protocols, progress in Monte Carlo techniques has permitted an improvement in accuracy in the determination of the absorbed dose. Nowadays, applications of the Monte Carlo method in medical physics span almost all topics, including radiation protection, diagnostic radiology, radiotherapy and nuclear medicine with an increasing interest in new applications such as nanoparticles technique, Boron Neutron Capture Therapy (BNCT), DNA damage and microdosimetry. Thanks to the rapid development of computational power, Monte Carlo based treatment plannings for radiation treatment are becoming feasible, too [13–15].

9.2.1 The Geant4 Simulation Toolkit

Geant4 (GEometry ANd Tracking) [16, 17] is one of the most widely used Monte Carlo toolkits to study particles interaction and transport in the matter. It is currently used in a large number of experiments and projects and in a variety of application domains, including high energy physics, astrophysics and space science, medical physics and radiation protection. The Geant4 diffusion is essentially due to its advanced functionalities in the geometrical description and to a wide and well-tested set of available physics models. The first product of the Geant *family* was released in 1974 at CERN (Geneva, CH) in order to simulate the interaction of high energy elementary particles with matter. It was limited to a restricted set of particles and simple detector geometry. In 1982 a new project started: its aim was to develop a completely new toolkit written in Fortran [18, 19]: it was called Geant3 and discontinued at the end of the nineties, even if many researcher still use it. It allowed to simulate large experimental apparata and the transport of high energy beams. The last member of the Geant4 family was born in 1998 when, after a first development phase, an Object Oriented, reusable and easy to maintain C++ toolkit was released. This new product was called Geant4.

Geant4 is currently developed by an international collaboration, constituted by about one hundred members from Europe, US Asia and Australia that release a new version of the code twice per year [20]. The public version of the Geant4 toolkit consists of a collection of C++ libraries containing all tools that the user has to include to develop his/her specific application. The object-oriented technology allows the easy and reusable development of distinct classes that can be used to simulate different aspects of a typical experiment: there are classes to construct the geometry, to define the materials, to model the source, the physics processes and so on. The public Geant4 distribution also contains a wide set of examples (divided in three main categories) representing the ideal users' starting point:

1. *Basic examples*: developed to illustrate the basic functionalities of the toolkit;
2. *Extended examples*: focused on many specific capabilities and domains on which Geant4 can be used;
3. *Advanced examples*: each describing a full experimental apparatus

Over the last years, the code evolved to address many requirements by a wide and growing community of Users and to encompass many physics applications and technological evolutions. The possibility to exploit the modern CPU multithreads has been, for example, recently introduced. Other different options for the future developments of Geant4 are also explored: GPUs [21] and accelerated processors offer, for example, great potentialities for speeding up intensive applications like those common in physics. Moreover, massively parallel computing and vectorization are being examined as a way to exploit available supercomputer capacity. In 2013 a new project was launched, named GeantV, to develop an all-particle transport simulation package with the broad objective of developing a tool based on Geant4 but several times faster to be used on CPU accelerators. A first alpha version of GeantV has been recently released [22].

9.2.2 Geant4 Use in Medical Physics

Thanks to its great flexibility and precision in terms of physics models, geometry and transport parameters, Geant4 allows the Users to simulate complex three-dimensional geometries, possibly importing Computers Aided Design (CAD) and/or Digital Imaging and COmmunications in Medicine (DICOM) images, routinely used in the medical physics community [23, 24]. Coupled with these advanced geometry capabilities, Geant4 offers several interesting features in terms of transport parameters,

Table 9.2 List of medical physics examples released inside the Geant4 distribution

Advanced examples	Extended examples
Brachytherapy	Dicom
Cell_irrdiation	dna
Human_Phantom	FanoCavity_1
Medical_linac	FanoCavity_2
IORT_Therapy	ElectronScattering_1
Hadrontherapy	ElectronScattering_2
GammaKnife	GammaTherapy

specifically introduced for dosimetric and micro/nano dosimetry studies. A particle can be transported until its kinetic energy goes down to approximately zero. Specific production thresholds of secondary particles (protons, electrons, positrons, and gamma) can be defined in order to optimize the dose depositions and the overall simulation precision. Several examples focused on medical applications are included in Geant4 distribution to guide the Users in the simulation of specific problems from the conventional radiotherapy to the Hadrontherapy and nuclear medicine applications (for diagnostic and therapy) and to the dosimetry at macro-, micro- and nano scale level. The medical physics examples in *Advanced* and *Extended* categories [25, 26] are listed in the Table 9.2.

Geant4 is currently able to simulate and model biological effects of ionizing radiation at the DNA scale [27, 28]. A tool extension, named Geant4-DNA and included in the low-energy Electromagnetic (EM) package, contains discrete EM physics models applicable to electrons, protons, neutral Hydrogen ions, Helium ions and its charged states in liquid water (that are the main component of biological media). This package is able to simulate all interactions step-by-step in order to precisely reconstruct track structures of ionizing particles at nanometre scale, thus allowing for a more precise understanding of damage at the DNA level [29].

9.3 Hadrontherapy: An Application for Clinical Passive Proton/Ion Beam Lines Simulation

9.3.1 Main Characteristics

Among the *Advanced Examples* released with the Geant4 distribution, Hadrontherapy was an application specifically developed for dosimetric and radiobiological studies with protons and ions beams [30, 31]. The application has been realized in the framework of a joint project between the INFN (Istituto Nazionale di Fisica Nucleare) and the Geant4 collaboration. The main purpose of this project, started in 2003, was the development of an open source Monte Carlo application able to reproduce each element of the CATANA (Centro di AdroTerapia ed Applicazioni Nucleari Avanzate) [32] beam line, the first Italian protontherapy facility dedicated to the treatment ocular melanomas. At the end of 2004, the first version of Hadrontherapy was released. Over the past several years, the application has been improved in order to get more flexibility and additional capabilities. Specific tools were added to permit the collection of information on the energy spectra of primary and secondary particles. The primary event generator has been optimized, with the introduction of a new approach based on phase space, allowing to run fast simulations also in the case of complex sources. Hadrontherapy also provides the possibility to activate physics lists via simple macro commands. Recently, a set of classes specifically devoted to the calculation of biological quantities: dose, averaged LET-dose, RBE, Survival Fraction and biological dose has been introduced [33]. The geometrical module of Hadrontherapy is divided in two main independent blocks, described by two different classes: a block is delegated the simulation of the beamlines geometry, the second is reserved to the simulation of the voxelized detector immersed in a water box. The detector is a specific volume (whose dimensions, position and material can be easily changed by macro commands) permitting the scoring of several quantities of interest such as deposited dose and particle fluences. Hadrontherapy allows the simulation of three passive proton/ion beamlines, including all necessary transport elements: scattering and modulation systems for spatial and energy distribution beam definition, collimators, transmission detectors and detectors for dose distribution measurements. In particular Hadrontherapy allows the precise simulation of three INFN beam lines:

- CATANA ocular proton therapy facility (INFN-LNS Catania, I);
- The Zero degree multidisciplinary facility (INFN-LNS Catania, I);
- TIFPA multidisciplinary beam line (INFN-TIFPA Trento, I) [34].

9.3.2 Reference Physics Lists

Since the release 10.0, Geant4 provides nine reference *physics lists* (classes containing a complete set of physics models and dedicated to specific applications) [35, 36]. Hadrontherapy provides three custom physics lists suggested for medical physics applications (see Table 9.3).

The electromagnetic interactions are simulated using the G4EmStandardPhysics_option4 builder, which implements a condensed-history algorithm based on the Bethe-Bloch energy loss formula [37]. This physics constructor was designed for applications for which high accuracy of electrons, hadrons and ion tracking is required. It includes the most accurate standard and low-energy models and it is suggested for the simulations focused on medical physics application. Hadronic interactions are reproduced with models implemented in the QGSP_BIC and QGSP_BIC_HP constructor, in which Geant4 native pre-equilibrium and de-excitation models are used as low energy stages of the Binary Cascade model for protons, neutrons and ions [38]. The builder QGSP_BIC_HP uses the addition data driven high precision neutron package (PartcleHP) to transport neutrons and light charged particle below 20 MeV down to thermal energies.

Table 9.3 Reference physics lists options available inside the Hadrontherapy application

	HADRONTHERAPY_1	HADRONTHERAPY_2	HADRONTHERAPY_3
Electromagnetic	standard_opt4	standard_opt4	standard_opt4
	G4EmExstraPhysics	G4EmExstraPhysics	–
	–	–	G4EmLivermorePhysics
Hadronic	G4DecayPhysics	G4DecayPhysics	G4DecayPhysics
	G4RadioactiveDecayPhysics	G4RadioactiveDecayPhysics	–
	G4IonBinaryCascadePhysics	–	G4IonBinaryCascadePhysics
	G4HadronElasticPhysicsHP	G4HadronElasticPhysicsHP	G4HadronElasticPhysics
	G4StoppingPhysics	G4StoppingPhysics	G4StoppingPhysics
	G4HadronPhysicsQGSP_BIC_HP	G4HadronPhysicsQGSP_BIC_HP	G4HadronPhysicsQGSP_BIC
	G4NeutronTrackingCut	–	G4NeutronTrackingCut

9.3.3   Primary Event Generator

The Hadrontherapy particle source module was planned to provide Users' with two choices: a standard random event primary generator and a phase space file, which contains data related to the events (position, direction of propagation, kinetic energy and particle type). The use of the phase space also allows to reduce the computing time without affecting the results accuracy. The phase space approach permits to divide the whole simulation into two steps. During the first step, the information of all particles hitting a scoring plane are recorded in a file from which they are recalled during the second step of the computation. This approach reduces the computing time considerably because the same primary data can be used in different simulations.

9.4   The CATANA Beamline and Its Complete Simulation

9.4.1   Set-Up of the Beamline

CATANA (Centro di AdroTerapia ed Applicazioni Nucleari Avanzate) was the first Italian protontherapy facility and in operation since 2002. The beamline is fully simulated inside the Hadrontherapy Geant4 application (Fig. 9.1). A picture of the real beam line is displayed in Fig. 9.2. Accelerated protons exit in air through a 50 μm kapton window. Range shifters and range modulators are positioned downstream the scattering system. Two transmission monitor ionization chambers provide the on-line control of the dose delivered to the patient. In order to monitor the beam stability during the treatment, a micro strip ionisation chambers detector [39] is placed at the end of the beamline. The beamline ends with a final (50 cm long, 36 mm in diameter) brass collimator, defining the final emittance of the beam at patient location.

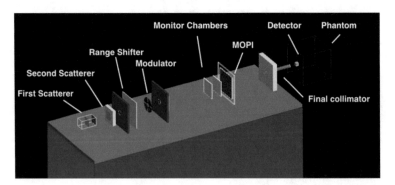

Fig. 9.1 The CATANA beamline geometry implemented inside Hadrontherapy application

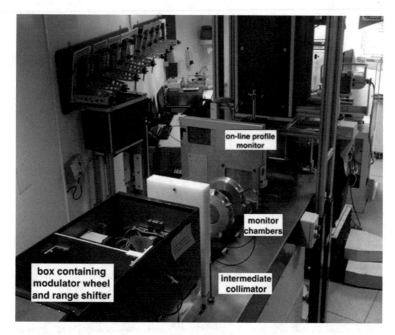

on-line profile
monitor

monitor
chambers

box containing
modulator wheel
and range shifter

intermediate
collimator

Fig. 9.2 Photo of the CATANA protontherapy beamline

9.4.2 Design of a Scattering System: The Role of Monte Carlo

Lateral and energy beam spread is obtained in CATANA by interposing static systems (scattering foils, plastic modulator wheels) along the beam path. Different approaches can be adopted to spread the accelerated beam laterally.

The simplest scattering system consists of a single scattering foil able to spread the beam into a Gaussian-like profile [40]. A collimator then fixes the beam dimensions. The scattering foil is generally made of high-Z materials (tantalum, gold, etc.) to obtain the maximum scattering minimising the energy losses (Fig. 9.3).

The efficiency of this scattering system is very low: spreading is limited to small fields of the order of millimeters [41]. A better efficiency can be achieved by using a double-scattering system: the first scatterer spreads the beam that then reaches the second scatterer. By inserting an occluding central stopper between the two scatterer a wider uniformity in the lateral dose distribution can be easily achieved. A layout of a double-scattering system is reported in Fig. 9.4.

The CATANA protontherapy facility [32] adopts such kind of system to obtain a lateral dose profile in accordance with the clinical ocular melanoma treatment prescriptions. Specifically, the double scattering system consists of two tantalum foils of 15 μm and 25 μm, respectively. The brass central stopper is 4 mm in diameter and 7 mm high. The first foil is located in the vacuum region of the beam transport line, about 20 cm upstream the in-air section, while the second is located in air, 15 cm

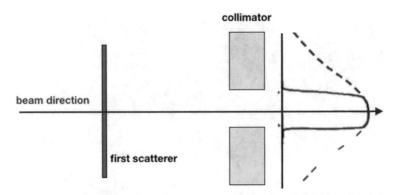

Fig. 9.3 Scattering system scheme based on a single scattering foil

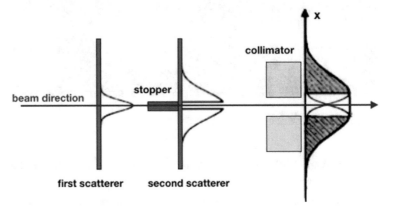

Fig. 9.4 Scattering system layout based on a double scattering foil with a central stopper

beyond the 15 μm kapton window separating the vacuum from the air section. The scattering system has been completely simulated and inserted into the Hadrontherapy application. The comparison of the experimental results and simulated data is reported in the Fig. 9.5.

9.4.3 Energy Modulation System

Energy modulation is achieved by means of PMMA modulator wheels. A modulator wheel is made of different steps of varying thickness able to reproduce pristine peaks of different energies to finally achieve a Spread-Out Bragg Peak (SOBP) [42]. Using analytical calculations based on the power-law approximation of the proton stopping power, it is possible to describe the proton Bragg peaks and calculate the optimal weights for a SOBP [43]. The Hadrontherapy application provides a dedicated class for the simulation of the CATANA modulator wheels. Users can

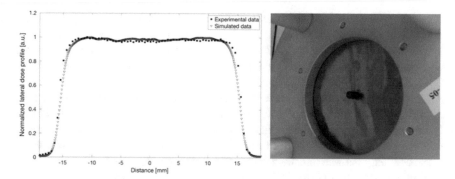

Fig. 9.5 Left: experimental lateral dose profile distribution (circles) of a 62 MeV proton beam and the corresponding simulation dose (triangles). Right: second scatterer with its stopper of the system installed at the CATANA facility

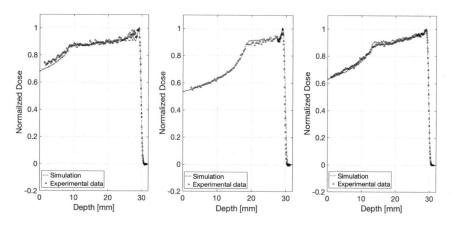

Fig. 9.6 Experimental and simulated Spread Out Bragg Peak obtained by using three different modulator wheels

change the modulation region according to the different geometrical simulation of the available wheels. An example of the available SOBPs that can be reproduced in Hadrontherapy is reported in Fig. 9.6.

9.5 Monte Carlo for the Extimation of Radiobiological Relevant Parameters

9.5.1 LET (Linear Energy Transfer) Calculation

Monte Carlo simulations offer a very powerful solution to obtain local energy spectra in a given geometry making use of the information retrieved step-by-step along particles track. Geant4 hence permits a precise calculation of the ratio between the total energy deposited and the total track lengths from all primary charged particles interacting with a given material, this information begin necessary for a precise and reliable estimation of the LET (Linear Energy Transfer) of the radiation in the considered medium. Today many approaches to compute the LET using a Monte Carlo code are available [44–47].

A completely new module, as respect what reported in [48], to compute the average dose-LET $\bar{L}_D$ and average track-LET $\bar{L}_T$ has been recently included inside the Hadrontherapy application. The new algorithms have been developed in order to produce results independent from the simulation transport parameters such voxel size, secondary particle threshold (production cuts) and step length.

Both algorithms (for dose-LET and track-LET calculation) make use of a method implemented in the Geant4 kernel and belonging to the G4EmCalculator class; the method converts the energy of charged particles to unrestricted LET, thus permitting the calculation of averaged dose-LET weighted by the electronic energy loss ε_{sn} using the formula:

$$\bar{L}_D = \frac{\sum_{n=1}^{N} \sum_{s=1}^{S_n} L_{sn} \cdot \varepsilon_{sn}}{\sum_{s=1}^{S_n} L_{sn} \varepsilon_{sn}} \tag{9.1}$$

$$\bar{L}_T = \frac{\sum_{n=1}^{N} \sum_{s=1}^{S_n} L_{sn} \cdot l_{sn}}{\sum_{s=1}^{S_n} L_{sn} l_{sn}} \tag{9.2}$$

Figure 9.7 shows the weighted dose-LET and track-LET derived using the (9.1) and (9.2) calculated by considering the contribute of the primary particles only (dash and dotted-dash curves) and the concurrent effect of the complex spectrum of the

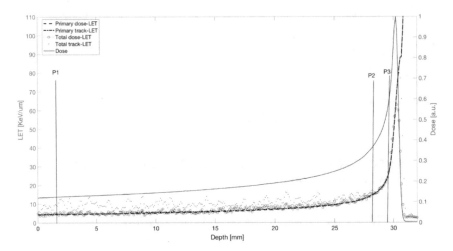

Fig. 9.7 Weighted dose and track-LET for a 60 AMeV 4He beam calculated using formula (9.1) and (9.2) for the primary beam and also taking into account the spectrum of secondary generated particles. 4He beam simulated dose distribution is also reported

secondary particles generated by the primary beam iteractions (dots and dots with the open circles respectively).

9.5.2 RBE (Relative Biological Effectiveness) Calculation

One of the most important parameters to describe the biological damage of an incident ion beam is the Relative Biological Effectiveness (RBE) [49]. Analytical calculations and Monte Carlo simulations (or their combination) represent nowadays one of the best approaches for RBE evaluation [50–53]. A dedicated class to compute RBE which Geant4 with radiobiological models have been included in the Hadrontherapy application and will be soon available to Users. The mixed field method, based on the alpha and beta parameters computed step-by-step is adopted [54] in this class. Pre-compiled Look Up Tables (LUTs) generated using the LEM model are coupled to the Geant4 simulation. Specifically, the LUTs have been generated using the version II and III of the LEM model containing the α_D and β_D values for a given monoenergetic ions (from $Z = 1$ to $Z = 8$) and cell type, and as function of ion LETs and energy [55].

Hadrontherapy was used to calculate RBE values for U87 human glioma cells irradiated with alpha particles and to compare results with those measured and calculated in the case of proton irradiation [56]. The simulation considers a tissue material, divided in small cubic voxels and irradiated with a given charged beam. Monte Carlo simulations can retrieve for each voxel and for each particle step the particle type, its kinetic energy, step-length, its energy loss. The combination of these quantities is the finally LET values. The kinetic energy and LETs of any primary ion and secondary in each voxel are calculated and used to perform a direct linear interpolation of the LUTs and to obtain a weighted alpha and beta value, accordingly to the formulas:

$$\langle \alpha_D \rangle = \frac{\sum_i \alpha_{D_i} \cdot D_i}{\sum_i D_i} \tag{9.3}$$

$$\langle \beta_D \rangle = \left(\frac{\sum_i \sqrt{\beta_{D_i}} \cdot D_i}{\sum_i D_i}\right)^2 \tag{9.4}$$

where $\alpha_D i$ and $\beta_D i$, refer to a specific radiation type i with a released dose D_i. In Fig. 9.8 the experimental survival fraction obtained irradiating U87 radioresistant human glioma cells at three different depths along a monochromatic proton beam of 62 MeV is reported. The corresponding simulated survival curve is also reported. The light-gray band represents the most probable region ($1.07\,\sigma$) where the experimental curves could fall on the basis of the measured error on α_D and β_D. Survival curve calculated using LEM (open squares) is also reported. More details on the experiment performed at the CATANA protontherapy facility of INFN-LNS can be found in Chaudhary et al. [56]. To shown the good reproducibility on the cell damage by

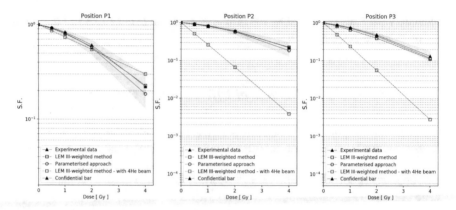

Fig. 9.8 Experimental survival curves for U87 cells irradiated at three different positions along a clinical proton pristine Bragg peak (see Fig. 9.9). The corresponding curves calculated using the Monte Carlo LEM weighted method (open square) is also reported in the two investigated cases: 62 MeV proton beam(black line) and 4He ion incident beam of 62 AMeV (blue line)

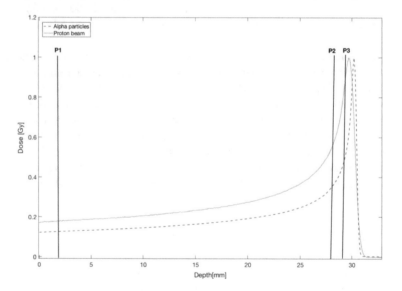

Fig. 9.9 Simulated dose released in water by an incident proton beam (blue line) and incident 4He (red line) in both cases with 62 AMeV of energy. The points P1, P2 and P3 corresponds to the investigated positions. The corresponding LET value is 1.2, 2.6 and 4.5 keV/μm in the case of incident proton beam and 13, 35 and 60 keV/μm for the He4 incident beam

using high-LET incident beam the Survival Curve with 62 AMeV incident 4He ions beam is also shown in Fig. 9.8.

The simulated dose released in water by both proton beam and 4He beam is shown in cite fig. Investigated positions along the Bragg curve are indicated with P1, P2 and P3.

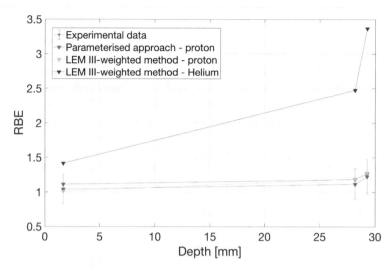

Fig. 9.10 Experimental RBE measured irradiating U87 cells in different positions along the Bragg curve (blue points). The calculated values by using Monte Carlo LEM-weighted method for protons (yellow points), for Helium (violet points), parametrized approach (red points) are reported

The dose-response curves well reflect the increase in RBE with increasing LET of the incident beam. In all investigated positions, the simulated curve is entirely of the experimental error bar. As shown in Fig. 9.8 alpha parameter increase proportionally to LET value. A comparison to the RBE calculated in the case of incident proton beam (low-LET particles) and 4He beam (high-LET particles) is shown in Fig. 9.10.

Accordingly, to the literature the RBE of the 4He beam is mostly higher than the corresponding effect due to the proton beam with the same incident energy. Along the Bragg curve differences in RBE becoming more entrenched consistent with a LET increasing.

9.6 Using Simulations with Laser-Driven Beams: The ELIMED Beamline at ELI-Beamlines

The acceleration of charged particle via ultra-intense and ultra-short laser pulses, has gathered a strong interest in the scientific community in the past years and it represents nowadays one of the most challenging topics in the relativistic laser-plasma interaction research. Indeed, it could represent a new path in particle acceleration and open new perspectives in multidisciplinary fields. Among many scenarios, one of the most interesting idea driving recent research activities, consists in setting up high intensity laser-target interaction experiments to accelerate ions for medical applications, with main motivation of reducing cost and size of acceleration, currently associated with big and complex acceleration facilities [57, 58].

Indeed, a development of more compact laser-based therapy centres could lead to a widespread availability of high-energy proton and carbon ion beams providing hadron therapy to a broader range of patients [57, 58].

However, to assume a realistic scenario where laser-accelerated particle beams are used for medical applications, several scientific and technological questions have to be answered and requirements to be fulfilled. Furthermore, the properties of laser driven proton bunches significantly differ from those available at conventional accelerators, both in terms of pulse duration and peak dose rate. Thus, many scientific and technical challenges must be solved, first to demonstrate the feasibility of unique applications with laser driven ion beams, and second to perform reliable and accurate physical and dosimetric characterization of such non-conventional beams, before starting any medical research and application. Different acceleration regimes have been experimentally investigated in the intensity range 10^{18}–10^{21} W/cm^2 in the so-called Target Normal Sheath Acceleration (TNSA) regime [59–61]. Acceleration through this mechanism employs relatively thin foils (about 1 μm), which are irradiated by an intense laser pulse (of typical duration from 30 fs to 1 ps). At peak intensities of the order of 10^{18} W/cm^2 hot electrons are generated in the laser-target interaction whose energy spectrum is strictly related to the laser intensity itself. The average energy of the electrons is typically of MeV order, e.g., their collisional range is much larger than the foil thickness. They can hence propagate to the target rear and can generate very high space-charge fields able to accelerate the protons contained in the target. The induced electric fields, in fact, are of the order of several teravolts per metre and, therefore, they can ionize atoms and rapidly accelerate ions normal to the initially unperturbed surface. Typical TNSA ion distribution shows a broad energy spread, exceeding 100%, much larger compared to the 0.1–1% energy spread typical of ion beams delivered by conventional accelerators, a wide angular distribution with an half-angle approaching 30^0 which is very different from the typical parallel beam accelerated by the conventional machine and a very high intensities per pulse, i.e. up to 10^{10}–10^{12} particles per bunch, as well as a very short temporal profile (ps) compared to 10^7–10^{10} particles/s of conventional clinical proton beams. Moreover, the cutoff energy value can be likely considered as a spectrum feature still strongly dependent to the shot-to-shot reproducibility and stability and up to now, the maximum proton energy obtained with a solid target in the TNSA regime is about 85 MeV [62].

These results are particularly promising along the pathway for achieving laser-driven ion beams matching the parameters required for different multidisciplinary applications, including the medical ones. Moreover, such improvements in the laser-driven source features will allow reaching better conditions for potential collection and transport of such kind of beams. Indeed, coupled to the investigations recently carried out on different target types, the development of new strategies and advanced techniques for transport, diagnostics and dosimetry of the optically accelerated particles represents a crucial step towards the clinical use of such non conventional beams and to achieve well-controlled high-energy beams with suitable and reproducible bunch parameters for medical applications. In this context, a collaboration between the INFN-LNS (Nuclear Physics Laboratory, Catania, Italy) and the ASCR-FZU

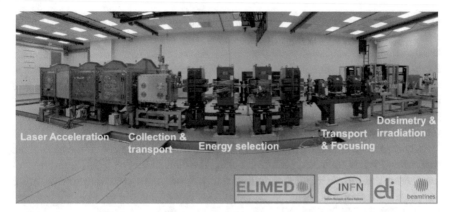

Fig. 9.11   The ELIMED line section installed at ELI-Beamlines (Doln Brezany, CZ) as part of the ELIMAIA beamline (July 2018)

(Institute of Physics of the Czech Academy of Science), has been established in 2011. The main aim of the collaboration, named ELIMED (ELI-Beamlines MEDical applications), is to demonstrate that high energy optically accelerated ion beams can be used for multidisciplinary applications, including the hadron therapy case, designing and assembling a complete transport beam-line provided with diagnostics and dosimetric sections that will also enable the Users to apply laser-driven ion beams in multidisciplinary fields. In 2012, ELI-Beamlines started the realization of the laser facility, where one of the experimental hall, will be dedicated to ion and proton acceleration and will host the ELIMED beamline. In 2014, a three-years contract has been signed between INFN-LNS and ELI-Beamlines to develop and realize the ELIMAIA beamline section dedicated to the collection, transport, diagnostics and dosimetry of laser-driven ion beams. This section, named ELIMED as the collaboration, was entirely developed by the LNS-INFN and has been delivered and assembled in the ELIMAIA experimental hall in July 2018. One of the purposes of the ELIMAIA beamline is to provide to the interested scientific community a user-oriented facility where accurate dosimetric measurements and radiobiology experiments can be performed. The technical solution proposed for the realization of the ELIMED beam line are described in [63]. The picture of the ELIMED beamlines installed at ELIMAIA (Doln Brezany, CZ) is shown in Fig. 9.11.

The beam transport line consists of an in vacuum section dedicated to the collection transport and selection of the optically accelerated particles. In particular, few cm downstream the target, a focusing system based on permanent magnet quadrupoles (PMQs) will be placed. A complete description of the designed system along with the study of the PMQs optics for different energies is given in [64]. The focusing system will be coupled to a selector system (ESS) dedicated to the beam selection in terms of species and energy. The ESS consists of a series of 4 C-shape electromagnet dipoles. The magnetic chicane is based on a fixed reference trajectory with a path length of about 3 m. According to the feasibility study results, such a solution will allow to deliver ions up to 60 MeV/n with an energy bandwidth, depending on the slit aperture, varying from 5% up to 20% at the highest energies and for the different species selected ensuring a rather good transmission efficiency, 10^6–10^{11} ions/pulse. At the end of the in vacuum beam-line, downstream the ESS, a set of conventional electromagnetic transport elements, two quadrupoles and two steering magnets, will allow refocusing of the selected beam and correcting for any possible misalignment. This last transport section will also allow providing a variable beam spot size between 0.1 and 10 mm.

A complete Monte Carlo simulation of the entire beamline and of the associated detectors [65] has been also performed using the Geant4 toolkit [16, 17, 66]. Moreover, when the system simulation will be ready, it will be used to study and optimize the particle transport at well defined positions. The evaluation of dose, fluence and particle distribution in the in-air section, will be performed, as well.

According to the beam transport simulation results, performed for the 60 MeV case with the beamline elements designed for ELIMAIA and considering a typical TNSA-like distribution with a cutoff energy of about 120 MeV and an angular divergence with a FWHM of 5 at 60 MeV, it is possible to deliver 60 MeV proton beam with a 20% energy spread with a rather uniform 10 × 10 mm spot size, beam divergence less than 0.5^0 and achieving a transmission efficiency of about 12%.

The simulation studies permitted us to estimate, in the worst conditions for the generated beams (biggest angular spread lowest expected particles number) the dose reaching the end of the beamline at each bunch. The value of 2 cGy per pulse, for

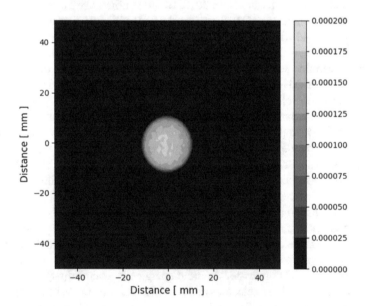

Fig. 9.12 Two dimensional dose distribution of a 30 MeV proton beam at the irradiation point and for one laser shot. The colours are representative of the released dose and values must be multiply by 100 to obtain the dose in Gy

60 MeV protons was found. This value, assuming a laser repetition rate of 1 Hz, would provide a pulsed proton beam with an average dose rate of about 1.2 Gy/min, which represents the minimal requirement for typical radiobiology experiments and is also promising considering the future possibility of running the PW laser available at ELI-Beamlines at a repetition rate of 10 Hz. In Fig. 9.12 is shown the simulated two dimensional dose distribution of a 30 MeV proton beam at the irradiation point and for one laser shot. Each colour represents a given value of dose delivered at the irradiation point but values must be multiplied by 100 to obtain a dose in Gy. This means that a maximum dose of 0.02 Gy per shot is reached for this simulation configuration. Such kind of information will be of capital importance for the schedule of the first biological irradiation at the ELIMED beamline and foreseen with 2020.

References

1. M.J. Berger, Monte Carlo calculation of the penetration and diffusion of fast charged particles, in *Methods in Computational Physics*, vol. I (Academic Press, New York, 1963), pp. 135–215
2. M.J. Berger, J.H. Hubbell, XCOM: photon cross sections on a personal computer, Technical Report NBSIR 87–3597 (National Institute of Standards and Technology, Gaithersburg, MD, 1987)
3. J. Seco, F. Verhaegen, *Monte Carlo Techniques in Radiation Therapy*. Series: Imaging in Medical Diagnosis and Therapy, 16 November 2016
4. N. Metropolis, *The Beginning of the Monte Carlo Method* (Los Alamos Science, 1987) (Special Issue), pp. 125–130
5. N. Metropolis, S. Ulam, Monte Carlo method. Am. Stat. Assoc. **44**, 335341 (1949)
6. H.A. Meyer (ed.), *Symposium on Monte Carlo Methods* (Wiley, New York, 1981)
7. G.C. de Buffon, Essai darithmtique morale, vol. 4. Supplment lHistoire Naturelle (1777)
8. G. Battistoni, S. Muraro, P.R. Sala, F. Cerutti, A. Ferrari, S. Roestler, A. Fasso, J. Ran, e FLUKA code: description and benchmarking, in *Proceedings of Hadronic Simulation Workshop, Fermilab*, vol. 68, September 2006
9. J. Briesmeister, MCNPA general purpose Monte Carlo code for neutron and photon transport, Version 3A. Los Alamos National Laboratory Report LA-7396-M (Los Alamos, NM, 1986)
10. M.J. Berger, ESTAR, PSTAR, and ASTAR: computer programs for calculating stopping-power and range tables for Electrons, Protons, and Helium Ions, Technical Report NBSIR 4999 (National Institute of Standards and Technology, Gaithersburg, MD, 1993)
11. J. Wulff, TOPAS/GEANT4 configuration for ionization chamber calculations in proton beams. Phys. Med. Biol. 106989 (2018)
12. A.Y. Chen, Y.W.H. Liu, R.J. Sheu, Radiation shielding evaluation of the BNCT treatment room at THOR: A TORT-coupled MCNP Monte Carlo simulation study. Appl. Radiat. Isot. **66**, 28–38 (2008)
13. B. Jones, S. Krishnan, S. Cho, Estimation of microscopic dose enhancement factor around gold nanoparticles by Monte Carlo calculations B. Med. Phys. **37**, 3809–3816 (2010)
14. S. Jan et al., GATE—GEANT4 application for tomographic emission: a simulation toolkit for PET and SPECT. Phys. Med. Biol. **49**(19), 4543–4561 (2004)
15. J. Allison et al., Recent developments in GEANT4. Nucl. Instrum. Methods A **835**, 186–225 (2016)
16. S. Agostinelli et al., GEANT4A simulation toolkit. Nucl. Instrum. Meth. A **506**, 250–303 (2003)
17. R. Brun, M. Hansroul, J.C. Lassalle, GEANT users guide. CERN Report DD/EE/82 (1982)

18. J. Allison et al., An application of the GEANT3 geometry package to the description of the opal detector. Comput. Phys. Commun. **47**, 55–74 (1987). CERN-EP/87-80
19. Official Geant4 website, http://geant4.web.cern.ch
20. S. Okada et al., GPU acceleration of Monte Carlo simulation at the cellular and DNA levels. Smart Innov. Syst. Technol. **45**, 323–332 (2015)
21. G. Amadio et al., The GeantV project: preparing the future of simulation, in *Journal of Physics: Conference Series 2015*, vol. 664 (2015), p. 072006
22. A. Kimura et al., DICOM interface and visualization tool for Geant4-based dose calculation, in *IEEE Nuclear Science Symposium Conference Record*, vol. 2 (2005), pp. 981–984
23. N.A. Graf, mesh2gdml: from CAD to Geant 4, in *IEEE Nuclear Science Symposium and Medical Imaging Conference Record* (2012), pp. 1011–1015
24. B. Caccia et al., MedLinac2: a GEANT4 based software package for radiotherapy. Annali dell'Istituto Superiore di Sanita **46**(2), 173–177 (2010)
25. S. Elles et al., Geant4 and Fano cavity test: where are we? in *Journal of Physics: Conference Series*, vol. 102, No. 1 (2008), p. 012009
26. S. Incerti, The Geant4-DNA project. Int. J. Model. Simul. Sci. Comput. **1**, 157–178 (2010)
27. M.A. Bernal et al., Track structure modeling in liquid water: a review of the Geant4-DNA very low energy extension of the Geant4 Monte Carlo simulation toolkit. Phys. Med. **31**, 861–874 (2015)
28. S. Guatelli et al., Review of Geant4-DNA applications for micro and nanoscale simulations. Phys. Medica **32**(10), 1187–1200 (2016)
29. G.A.P. Cirrone, G. Cuttone et al., Implementation of a new Monte Carlo GEANT4 simulation tool for the development of a proton therapy beam line and verification of the related dose distributions. IEEE Trans. Nucl. Sci. **52**, 1756–1758 (2005)
30. G.A.P. Cirrone, G. Cuttone et al., Hadrontherapy: A 4-based tool for proton/ion-therapy studies. Prog. Nucl. Sci. Technol. **2**, 207–212 (2011)
31. G.A.P. Cirrone et al., Clinical and research activities at the CATANA facility of INFN-LNS: from the conventional hadrontherapy to the laser-driven approach. Front. Oncol. (2017). https://doi.org/10.3389/fonc.2017.00223
32. G. Petringa et al., Development and analysis of the track-LET, dose-LET and RBE calculations with a therapeutical proton and ion beams using Geant4 Monte Carlo toolkit. Phys. Med. **42**(1), 9 (2017)
33. F. Tommasino et al., Proton beam characterization in the experimental room of the Trento Proton Therapy facility. Nucl. Instrum. Methods Phys. Res. A **869**, 1520 (2017)
34. I. Kyriakou et al., Improvements in Geant4 energy-loss model and effect on low-energy electron transport in liquid water. Med. Phys. **42**, 3870–3876 (2015)
35. H. Wright, M.H. Kelsey, The Geant4 Bertini cascade. Nucl. Instrum. Methods A **804**, 175–188 (2015)
36. L. Pandola, Validation of the Geant4 simulation of bremsstrahlung from think targets below 3 MeV. Nucl. Instrum. Methods B **350**, 41–48 (2015)
37. J. Apostolakis et al., Progress in hadronic physics modelling in Geant4, in *Journal of Physics: Conference Series*, vol. 160 (2009), p. 012073
38. N. Givehchi et al., Online monitor detector for the protontherapy beam at the INFN Laboratori Nazionali del Sud-Catania. Nucl. Instrum. Methods Phys. Res. Sect. A Accel. Spectrom. Detect. Assoc. Equip. **572**(3), 1094–1101 (2006)
39. E. Grussel et al., A general solution to charged particle beam flattening using an optimized dual scattering foil technique with application to proton therapy beams. Phys. Med. Biol. **50**(5), 7 (2005), pp. 755–767
40. Y. Takada, Dual-ring double scattering method for proton beam spreading. J. Appl. Phys. **33**(59) (1994), p. 353
41. H. Paganetti, *Proton Therapy Physics*, Series in Medical Physics and Biomedical Engineering (2012)
42. S. Bijian Jia et al., Designing a range modulator wheel to spread-out the Bragg peak for a passive proton therapy facility. Nucl. Instrum. Methods Phys. Res. Sect. A Accel. Spectrom. Detect. Assoc. Equip. **806**, 101–108 (2015)

43. A.M. Kellerer, D. Chmelevsky, Criteria for the applicability of LET. Radiat. Res. **63**, 226–234 (1975)
44. A.M. Kellerer, Fundamentals of microdosimetry, in *Dosimetry of Ionizing Radiation*, ed. by K.R. Kase, B.E. Bjarngard, F.H. Attix, vol. 1 (Academic Press, Orlando, 1985), pp. 77–162
45. F. Guan et al., Analysis of the track- and dose-averaged LET and LET spectra in proton therapy using the Geant4 Monte Carlo code. Med. Phys. **42**(11) (2015)
46. M.A. Cortes-Giraldo et al., A critical study of different Monte Carlo scoring methods of dose average linear-energy-transfer maps calculated in voxelized geometries irradiated with clinical proton beams. Phys. Med. Biol. **60**, 264569 (2015)
47. F. Romano et al., A Monte Carlo study for the calculation of the averaged linear energy transfer (LET) distributions for a clinical proton beam line and a radiobiological carbon ion beam line. Phys. Med. Biol. **59**, 2863–2882 (2014)
48. IAEA TRS 461 Relative Biological Effectiveness in Ion Beam Therapy, International Atomic Energy Agency, 2008
49. R.B. Hawkins, A microdosimetric-kinetic model of cell death from exposure to ionizing radiation of any LET, with experimental and clinical applications. Int. J. Radiat. Biol. **69**(6), 739–755 (2017)
50. M. Scholz, Track structure and the calculation of biological effects of heavy charged particles. Adv. Space Res. **18**, 5–14 (1996)
51. M. Scholz et al., Computation of cell survival in heavy ion beams for therapy, the model and its approximation. Radiat. Environ. Biophys. **36**, 59–66 (1997)
52. A.L. McNamara et al., A phenomenological relative biological effectiveness (RBE) model for proton therapy based on all published in vitro cell survival data. Phys. Med. Biol. **60**(21), 8399–8416 (2015)
53. T. Elsasser et al., Accuracy of the local effect model for the prediction of biologic effects of carbon ion beams in vitro and in vivo. Int. J. Radiat. Oncol. Biol. Phys. **71**, 866–872 (2008)
54. P. Chaudhary et al., Relative biological effectiveness variation along monoenergetic and modulated bragg peaks of a 62-MeV therapeutic proton beam: a preclinical assessment. Int. J. Radiat. Oncol. Biol. Phys. **90**(1), 27–35 (2014)
55. A. Mairani et al., The FLUKA Monte Carlo code coupled with the local effect model for biological calculations in carbon ion therapy. Phys. Med. Biol. **55**, 4273–4289 (2010)
56. S.V. Bulanov et al., Feasibility of using laser ion accelerators in proton therapy. Plasma Phys. Rep. **28**(5), 453–456 (2002)
57. S.V. Bulanov, J.J. Wilkens et al., Laser ion acceleration for hadron therapy. Rev. Top. Probl. **57**(12), 1149–1179 (2014)
58. S.P. Hatchett, C.G. Brown et al., Electron, photon and ion beams from the relativistic interaction of petawatt laser pulses with solid targets. Phys. Plasmas **7**, 2076–2082 (2000)
59. S.C. Wilks, A.B. Langdon et al., Energetic proton generation in ultra-intense laser-solid interactions. Phys. Plasmas **8**, 542–549 (2001)
60. A. Macchi, A. Sgattoni, S. Sinigardi, M. Borghesi, M. Passoni, Advanced strategies for ion acceleration using high-power lasers. Plasma Phys. Control. Fusion **55**, 124020 (2013)
61. F. Wagner, O. Deppert, C. Brabetz, P. Fiala et al., Maximum proton energy above 85 MeV from the relativistic interaction of laser pulses with micrometer thick CH_2 targets. Phys. Rev. Lett. **116**(20), 205002 (2016)
62. G.A.P. Cirrone, F. Romano et al., Design and status of the ELIMED beam line for laser-driven ion beams. Appl. Sci. (2015)
63. F. Schillaci, G.A.P. Cirrone et al., Design of the ELIMAIA ion collection system. J. Instrum. (2014)
64. F. Romano, A. Attili et al., Monte Carlo simulation for the ELIMED transport beamline, in *AIP Conference Proceedings*, vol. 1546 (2013)
65. S. Agostinelli et al., Geant4—a simulation toolkit. Nucl. Instrum. Methods Phys. Res. A **506**, 250–303 (2003)
66. D.W.O. Rogers, Fifty years of Monte Carlo simulations for medical physics. Phys. Med. Biol. **51**(13) (2006)

67. R.R. Wilson, Monte Carlo study of shower production. Phys. Rev. **86**, 261–269 (1952)
68. J. Allison et al., Geant4 developments and applications. IEEE Trans. Nucl. Sci. **53**(1), 270–278 (2006)

Chapter 10
Lectures About Intense Lasers: Amplification Process

Bruno LeGarrec

Abstract The laser lecture about intense lasers is dedicated to the basic process of stored energy and amplification and the way to achieve high intensity when considering the shortest possible pulse duration. The amplification course is essentially based on the semi-classical model that represents the interaction between electromagnetic radiation and an assembly of atoms. On the radiation side, we always start with Maxwell's equations to arrive at the Helmholtz propagation equation, whereas on the atomic side the system will be represented by the two-level model. From there we can extend this model to the case of amplification, gain and stored energy. Many useful quantum aspects will be mixed with this model but we will always try to strive for the greatest possible simplification in order to have the most user-friendly formulas possible and especially not to lose the underlying physical reality. For accessing high peak powers, we will give a short description of the different techniques that were used since the very beginning of laser operation in 1960 and we will show that a major step was made in 1985 when the chirped pulse amplification process (CPA) was demonstrated. This year 2018 was somehow the celebration of this major event when the Nobel prize in Physics was delivered to Gérard Mourou and Donna Strickland.

10.1 Introduction

Intense lasers are lasers delivering high intensity beams. Laser intensity is defined as the ratio of the laser power to the laser beam area. Laser power is simply laser energy divided by the laser pulse duration. Finally high intensity means high energy, short pulse duration and small beam area. Small beam area is possible in the vicinity of the focal spot and the larger the spectral bandwidth, the shorter the pulse duration. Petawatt lasers are in the run for more than a decade [1] and some authors are already promising exawatt and zettawatt applications [2]. At the same time lasers scientists are trying to extract more and more energy from the laser medium and

B. LeGarrec (✉)
LULI, Ecole Polytechnique, Route de Saclay, 91128 Palaiseau Cedex, France
e-mail: bruno.le-garrec@polytechnique.edu

© Springer Nature Switzerland AG 2019

L. A. Gizzi et al. (eds.), *Laser-Driven Sources of High Energy Particles and Radiation*,
Springer Proceedings in Physics 231, https://doi.org/10.1007/978-3-030-25850-4_10

recent developments have shown that the kilojoule level is at hand [3, 4]. In this lecture we will concentrate at first on energy; how to amplify it and how to store it and then we will describe the different techniques that were used to decrease the pulse duration in the laser process: from "natural" relaxed oscillation to chirped pulse amplification process (CPA) while going through Q-switch and Mode-locking.

The system of international units was chosen and a rigorous presentation of the various most useful equations will be made. Nevertheless, by convention of use one will use certain notations and certain "hybrid" units which are frequent in the documents and the jargon of the laser people. Light is an electromagnetic radiation, therefore a wave that propagates at high speed ($c = 3 \; 10^8$ m/s) will travel 300 000 km in 1 s and 30 cm in one nanosecond (1 ns $= 10^{-9}$ s). The laser domain extends from X-rays to infrared (10 μm).

This lecture will cover the black body radiation, the two-level system (Einstein coefficients, their relationship and their properties), the amplification process (Frantz and Nodvik model) and finally a short description of the basic principle of the different techniques for achieving short pulse duration. The laser lectures are intended for students and beginners.

10.2   The Black Body Radiation

The semi-classical model between electromagnetic radiation and an assembly of atoms starts from the formulation of the black body. At the end of the 19th century, physicists sought the answer to the following question: what is the law describing the spectral density of radiation emitted by a gas contained in a chamber at a given temperature? Various models of the time well explained the behaviour at long wavelengths (Wien's law) but not at short wavelengths: this is what was called the ultraviolet "catastrophe". Planck has found the solution by resorting to a mathematical device which he did not explain and which he baptized the unknown "h" (in German, the equivalent of our "x") [5]. This amounts to assuming that the spectrum of energy levels is not continuous but discrete. Bold hypothesis for the time; even Planck was very dissatisfied. In 1905, Einstein explained the photoelectric effect [6] and thus gave the answer to the unknown "h" of Planck's formula. He introduced the notion of "photon" energy $E = h\nu$. One photon at $\lambda = 1$ μm wavelength carries an energy $E = h\nu = hc/\lambda$. Numerical application gives [6,62 10^{-34} J s] * [3 10^8 m/s]/[10^{-6} m] $= 2 \; 10^{-19}$ J so roughly 1 eV. The energy spectral density $\rho(\nu) = U(\nu)dN(\nu)/V$ is equal to the product of the average number of photons per mode by the energy of a photon and the density of modes between ν and $\nu + d\nu$. This is Planck's formula of black body radiation (Fig. 10.1):

$$\rho(\nu)d\nu = h\nu \; \frac{1}{\exp(\frac{h\nu}{kT}) - 1} \; \frac{8\pi \nu^2}{c^3} \; d\nu \qquad (10.1)$$

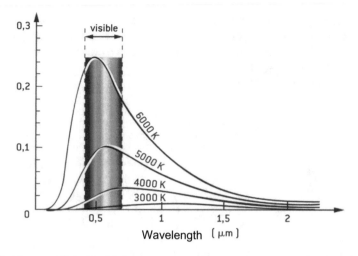

Fig. 10.1 The curve of the black body at different temperatures as a function of the wavelength

With $h\nu$, the photon energy; $1/[exp(h\nu/kT) - 1]$, the average number of photons per mode and $8\,\pi\nu^2/c^3$, the density of modes between ν and $\nu + d\nu$.

The energy spectral density varies as $\nu^3/[exp(h\nu/kT) - 1]$ while the spectral density per unit wavelength varies as $\lambda^{-5}/[exp(hc/\lambda kT) - 1]$. The curve of the black body always goes through a maximum which is governed by the law of Wien:

$$\lambda max.T \approx 3000 \text{ K } \mu m \tag{10.2}$$

In photometry, this temperature is called the luminance temperature and in plasma physics it is called the radiative temperature. This formula is very practical. A black body with a temperature of 6000 K emits a radiation whose maximum is at 0.5 μm, thus visible whereas a black body at 300 K will emit a radiation whose maximum is at 10 μm, therefore in the infrared.

10.3 The Two-Level System

Systems with 2, 3, 4 levels or more are reduced to a 2-level system for the sake of simplicity, but do not forget the degeneracy or the width of the levels. The degeneracy $g = 2\,m + 1$ is the multiplicity related to the number of sub-levels of kinetic moment given: orbital kinetics L, m_L, total $J = L + S$, J, m_J, or $F = J + I$, F, m_F. This degeneracy can be seen by the presence of an external static field (Stark effect, Zeeman effect). The width of the level depends on several phenomena that are classified into two categories. Homogeneous broadening is related to the lifetime of the atomic system (free atoms, electrons of the crystal lattice, molecules) while inhomogeneous broadening represents the Doppler effect of moving atoms and molecules.

The homogeneous width is characterized by a Lorentzian function and the inhomogeneous width by a Gaussian function, but one can meet more complex phenomena where the shape of the line is the convolution of the two preceding phenomena: one speaks then of Voigt's profile.

Lorentzian profile:

$$g(\omega) = \frac{\Delta}{2\pi \left[(\omega - \omega_0)^2 + \frac{\Delta^2}{4} \right]} \tag{10.3}$$

Gaussian profile:

$$g(\omega) = \frac{2}{\Delta} \sqrt{\frac{ln2}{\pi}} Exp\left(-4ln2 \left(\frac{\omega - \omega_0}{\Delta} \right)^2 \right) \tag{10.4}$$

We will generally write the full width at half maximum (FWHM) of general shape as the sum of radiative and the non-radiative contributions: $\Delta = 1/T_{rad} + 1/T_{non\,rad}$ with $\omega = 2\pi \nu$ is the pulsation of the wave whose width is well defined as the inverse of a time: $\Delta \omega = 1/T$. Furthermore:

$$\int g(\omega) d\omega = 1 \tag{10.5}$$

10.3.1 The Einstein Coefficients

At first, let us summarize what was known at the end of the 19th century about the basic processes of absorption and emission. In a typical light absorption experiment, the intensity transmitted by the material is lower than the incident intensity. Apart from the Fresnel losses (reflection on the faces), the difference of intensity is absorbed by the material. It was known that instantaneously, light can be emitted by the material (not necessarily at the same wavelength) in all directions and for a given time. In some cases, this emission persisted even when the exciter radiation was no longer present. We were talking about fluorescence and phosphorescence (Fig. 10.2).

In his 1917 article "On quantum theory of radiation", Einstein introduced the concept of amplification of radiation through the notion of stimulated emission [7]. After discovering the photoelectric effect that provided a justification for Planck's famous constant h, Einstein will introduce an additional phenomenon in the absorption-emission process in order to make this process more symmetrical. These works are in agreement with those of Niels Bohr (1913) on the planetary model of the hydrogen atom.

The three basic mechanisms are:

Absorption (superscript "a"):

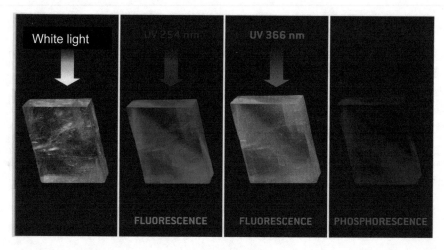

Fig. 10.2 A calcite crystal placed under a UV lamp emits fluorescence whose color is not the same as the excitation wavelength (254 or 366 nm). After suppression of UV radiation, the emission of light persists: it is phosphorescence

$$dN_0^a = -N_0 \, B_{0j} \, \rho(v) \, dt \tag{10.6}$$

Stimulated or induced emission (superscript "ei"):

$$dN_0^{ei} = +N_j \, B_{j0} \, \rho(v) \, dt \tag{10.7}$$

Spontaneous emission (superscript "es") (Fig. 10.3):

$$dN_0^{es} = +N_j \, A_{j0} \, dt \tag{10.8}$$

We consider an isolated or closed system where the total population of the levels is constant:

Fig. 10.3 A two-level system with the three basic mechanisms

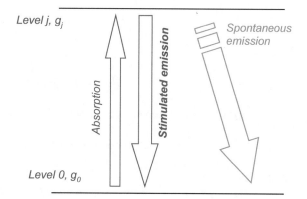

$$N = N_0 + N_j = cste \tag{10.9}$$

N_j is the population (or population density) of level j. We write the transition probability between levels:

$$P_{0 \to j} = B_{0j}\, \rho(v)\, dt \tag{10.10}$$

The balance population is written:

$$\Delta N_0 = dN_0^a + dN_0^{ei} + dN_0^{es} = [(N_j B_{j0} - N_0 B_{0j})\rho(v) + N_j A_{j0}]\, dt \tag{10.11}$$

And also:

$$\Delta N_j = -\Delta N_0 = [(N_0 B_{0j} - N_j B_{j0})\rho(v) - N_j A_{j0}]\, dt \tag{10.12}$$

We will write generally dN_j/dt, dN_0/dt.

10.3.2 Relationship Between Einstein Coefficients

It is known that at thermodynamic equilibrium, populations follow Boltzmann's law:

$$N_j \propto g_j\, exp - E_j/kT \tag{10.13}$$

and for a closed system, thus isolated, the absorption is equal to the sum of the emissions (induced and spontaneous). Hence:

$$g_0 B_{0j} = g_j B_{j0} \text{ and } A_{j0}/B_{j0} = 8\pi h v^3/c^3 \tag{10.14}$$

with N_j the population (or n_j the population density) of level j.
  On the other hand,

$$\rho(v)\, dv = \rho(\omega)\, d\omega \tag{10.15}$$

therefore

$$\rho(\omega) = \rho(v)/2\pi \tag{10.16}$$

$B_{j0}\, \rho(v)/A_{j0} = n(v)$ is the number of photons per mode and we can write:

$$\rho(v) = (8\pi v^2/c^3)\, h v n(v) \tag{10.17}$$

where

$$n(v) = 1/Exp\,(hv/kT) - 1) \tag{10.18}$$

We can conclude that in the optical domain where $hv \gg kT$ then the stimulated over spontaneous ratio $\approx Exp(-hv)/kT$ is favourable to the spontaneous regime while in the thermal domain $hv \approx kT$ then the stimulated over spontaneous ratio $\approx kT/hv$ and therefore favourable to the stimulated regime. In a medium of index n, we define the intensity in the laser sense as $I = c\rho(v)/n$ and the populations are written:

$$dN_j/dt = \left(N_0\,B_{0j} - N_j\,B_{j0}\right)\rho(v) - N_j\,A_{j0} \tag{10.19}$$

$$dN_j/dt = (n\,I(v)\,/c)\,A_{j0}\left(c^3/n^3\right)/\,8\pi h v^3\left(N_0 g_j/g_0 - N_j\right) - N_j A_{j0} \tag{10.20}$$

$$dN_j/dt = -\,A_{j0}(\lambda^2/8\,\pi v^2)\,(I(v)/hv)\left(N_j - N_0\,g_j/g_0\right) - N_j A_{j0} \tag{10.21}$$

10.3.3 Linewidth Broadening

In the general case, we have $\rho(v)$ and $g(v)$, and we replace A by $A' = Ag(v)$ and B by $B' = Bg(v)$. Two extreme cases are to be considered.

Either we have a narrow transition in front of the radiation density $\rho(v)$ and the linewidth is assimilated to a Dirac:

$$g(v) = \delta(v - v_0) \tag{10.22}$$

or we have a wide transition in front of the radiation density and the radiation is assimilated to a Dirac:

$$\rho(v) = I(v)/c = I_0 g(v - v_0)/c \tag{10.23}$$

10.3.4 Properties of the Relationship Between Einstein's Coefficients

Spontaneous emission is isotropic (photons are emitted in a random spatial direction) and un-polarized in general, except in specific cases (for example, if Brewster incidence dioptres are present). The stimulated emission has characteristic properties: the stimulated radiation has the same direction and the same polarization as the incident radiation. We can make an assessment of an intensity point of view and lead to the notions of gain and amplification.

10.4 Amplification

The intensity balance is written:

$$\Delta I = I_{transmitted} + I_{spontaneous} - I_{incident} \tag{10.24}$$

as a function of Einstein's coefficients when considering a 2-level system (level 1 and level 2) of thickness Δz and spontaneous emission inside a solid angle $d\Omega$:

$$\Delta I = h\nu B_{21} I(\nu)/c\, g(\nu) N_2 \Delta z - h\nu B_{12} I(\nu)/c\, g(\nu) N_1 \Delta z + h\nu A_{21} \Delta n\, g(\nu) N_2 \Delta z d\Omega/4\pi \tag{10.25}$$

$$\Delta I/\Delta z = h\nu B_{21}(N_2 - N_1 g_2/g_1) g(\nu) I(\nu) + h\nu A_{21} \Delta n\, g(\nu) \Delta z d\Omega/4\pi \tag{10.26}$$

There is gain whenever: (Fig. 10.4)

$$N_2 > N_1 g_2/g_1 \tag{10.27}$$

And this is called population inversion. Because of the Boltzmann law, such a population inversion is not possible in a 2-level system. Only 3-level and 4-level systems are suitable for population inversion.

The gain term is written:

$$dI(\nu)/dz = A_{21}(\lambda^2/8\pi \nu^2) g(\nu)(N_2 - N_1 g_2/g_1) I(\nu) = g_0(\nu) I(\nu) \tag{10.28}$$

$g_0(\nu)$ or $\gamma_0(\nu)$ is called *small signal gain* because $I_{incident}$ is small when compared with a characteristic intensity $I_{saturation}$ (that will be defined later on). The first part of $g_0(\nu)$ is the transition cross section. There are the emission cross section and the absorption cross section.

$$\sigma_{se} = A_{21}(\lambda^2/8\pi \nu^2) g(\nu) \tag{10.29}$$

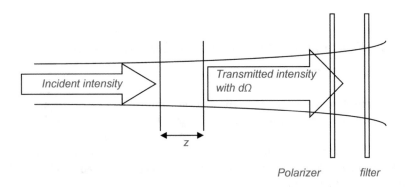

Fig. 10.4 Intensity balance during the absorption and emission processes

$$\sigma_{ab} = A_{21}(\lambda^2/8\pi\nu^2)\,g(\nu)\,g_2/g_1 \tag{10.30}$$

Finally with $\Delta N = N_2 - N_1\,g_2/g_1$, (10.27) can be simplified to:

$$g_0(\nu) = \sigma_{se}\Delta N \tag{10.31}$$

When it is possible to integrate $dI(\nu)/dz$ over z:

$$I\nu(z) = I\nu(0)\,Exp[g_0(\nu)\,z] \tag{10.32}$$

and

$$G_0(\nu) = Exp[g_0(\nu)z] = I\nu(z)/I\nu(0) \tag{10.33}$$

Equation (10.32) is the classical gain definition.

The characteristic value is called the saturation intensity I_{sat} (W/cm^2) but the most commonly used term is the saturation fluence F_{sat} (J/cm^2):

$$I_{sat} = h\nu/\sigma\tau \qquad F_{sat} = h\nu/\sigma \tag{10.34}$$

with τ, the pulse duration and σ, the emission cross section.

The saturation fluence depends only on the laser medium and its typical range is 0.1–10 J/cm^2. A low saturation fluence means that the stored energy will be low but the gain will be high because of the emission cross section. On the other hand a large saturation fluence is good for energy storage but the gain will be low. From this point of view, Nd:glass laser medium has a saturation fluence around 5 J/cm^2 and is considered as an ideal amplifying medium.

10.4.1 Gain Saturation and the Amplification Process According to Frantz and Nodvik

There is gain $g = \sigma\,\Delta n$ and this can be written either in intensity:

$$I_{out} = I_{in}Exp(g.l) \tag{10.35}$$

or fluence:

$$F_{out} = F_{in}Exp(g.l) \tag{10.36}$$

But what is happening when the amplification length is increasing indefinitely? Let's split the medium into «n» slices (equivalent to «n» amplifiers following each other's). Whenever 1 photon is emitted from the upper energy level E$_2$ to the lower energy level E_1, then the population of E_2 decreases of 1 unit and the one of E_1 increases of 1 unit because $g = \sigma \Delta n$, and finally, the extracted energy is:

$$\Delta E = \Delta N \, h\nu \tag{10.37}$$

The more I amplify, the less gain I get because ΔN is decreasing along the amplification process. Starting from Maxwell's equations, we can get the propagation equation «Helmoltz» type and the rate equations that can be simplified to:

$$\frac{\partial I}{\partial Z} = \sigma I \Delta N \tag{10.38}$$

and

$$\frac{\partial \Delta N}{\partial t} = -\frac{2\sigma I}{h\nu} \Delta N \tag{10.39}$$

This simplification is neither straightforward nor easy and can be found in references [9, 10]. Equations (10.38), (10.39) are known as the Frantz and Nodvik's equations of a two-level system [8]: this system has no analytical solution. Frantz et Nodvik are using transformation variables such that the space variable «z» becomes «z/c» and the time variable «t» becomes «t − z/c». The final Frantz and Nodvik solution can be written:

$$F_{out} = F_{sat} Ln(1 + Exp(gl))\left[Exp(F_{in}/F_{sat}) - 1\right] \tag{10.40}$$

It can be shown that:

$$Exp\left(\frac{F_{out}}{F_{sat}}\right) - 1 = Exp(gl)\left[Exp(F_{in}/F_{sat}) - 1\right] \tag{10.41}$$

and:

$$F_{in} \ll F_{sat} : F_{out}/F_{in} = Exp(gl) \tag{10.42}$$

$$F_{in} \gg F_{sat} : F_{out} - F_{in} = gl F_{sat} \,or\, \Delta E = gl F_{sat} \Delta S \tag{10.43}$$

The residual gain in one amplification slice "i":

$$g_i l = -Ln\left[1 - \left(Exp(-F_{in(i-1)}/F_{sat})\right)\left(1 - Exp(-g_{i-1}l)\right)\right] \tag{10.44}$$

From these two formulae, a recursive calculation in an EXCEL type spreadsheet is straight forward.

10.4.2 Energetic and Temporal Behaviour

As an example we have considered the output fluence and the residual gain in a simple "ideal" case of a long laser rod and we have decided to split it in equal slices. Saturation fluence is 4 J/cm², initial small signal gain is 0.05 cm^{-1} and we are considering an input fluence equal to 0.04, 0.4 and 4 J/cm² ($F_{sat}/100$, $F_{sat}/10$ and F_{sat}).

At low fluence the curve has an exponential shape (according to 10.42) while at high fluence it is close to a straight line (10.43). The length of the laser rod was chosen such that the output fluence will be much greater than the saturation fluence. When considering 2 cm slice and 10 cm × 10 cm area, the stored energy as given by (10.43) is [0.05 cm^{-1}] * [130 cm] * [100 cm²] * [4 J/cm²] = 2600 J. The extracted energy is simply the difference between the output energy and the input energy and at an input fluence equal to 0.04, 0.4 and 4 J/cm² it is respectively equal to 812, 1665 and 2417 J.

Equation (10.43) gives the stored energy. This is a very important parameter once you have chosen the amplification medium whose basic parameter is the saturation fluence. The next step for designing your amplifier is to decide the dimensions of your amplifying module: beam area and propagation length. From the conditions and results of Fig. 10.5, you know how much energy can be extracted at least in a single pass amplification scheme (Fig. 10.6).

At low input fluence, the residual gain remains close to the initial value up to the middle of the laser rod (black curve) while at fluence equal to the saturation fluence, it decreases quickly. In single pass amplification a lot of stored energy is not

Fig. 10.5 Output fluence after each slice when the input fluence on the first slice is 0.04 J/cm² (black curve), 0.4 J/cm² (blue curve) and 4 J/cm²(red curve)

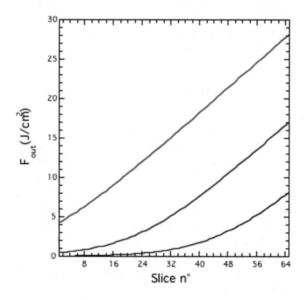

Fig. 10.6 Residual gain after each slice when the input fluence on the first slice is 0.04 J/cm² (black curve), 0.4 J/cm² (blue curve) and 4 J/cm²(red curve)

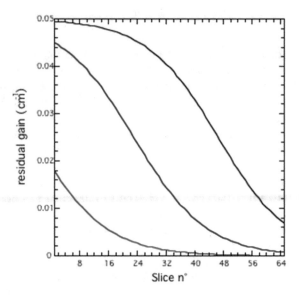

extracted at low fluence. Multiple pass amplification is a much better solution (see next paragraph).

For a given z (let us assume an infinitely thin slice of amplifier) at $t = 0$, the gain is g_0. At any time $t > 0$, the gain is lower since population inversion has been already used according to (10.37). So a square input pulse turns into a decreasing exponential pulse shape. Consequently, if I want a square pulse at the laser output, I have to generate an exponentially increasing input pulse. For Lorentzian or Gaussian shaped pulses, the front edge is more amplified than the last part of the pulse and very quickly the pulse is no longer symmetric.

10.4.3 Multiple Pass Amplification

For improving the extracted energy, it is better to use a multiple pass scheme where the amplified laser beam is travelling back and forth through the same laser medium. An example of a double-pass amplifier is given below and it is possible to increase the number of passes. Double-pass amplifiers are easy to build and most commonly used multiple-pass amplifiers have 4 passes. When amplifying from a very low fluence level and that a large number of passes is necessary, it is possible to use regenerative amplification as discussed by Murray and Lowdermilk in [11, 12] (Figs. 10.7, 10.8).

During the first pass, the gain is decreasing quickly but the residual gain remains high at the beginning of the laser rod (black curve) while during the second pass, the residual gain is small and almost flat which means that all stored energy has been extracted. With a laser rod half size of the one considered above, extraction efficiency in double pass is much better than is single pass. The jump in the residual

Fig. 10.7 Output fluence
after each slice in a double
pass design when the input
fluence on the first slice is
1 J/cm². Fist pass (black
curve) and second pass (blue
curve)

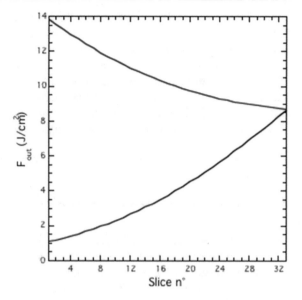

Fig. 10.8 Residual gain
after each slice in a double
pass design when the input
fluence on the first slice is
1 J/cm². Fist pass (black
curve) and second pass (blue
curve)

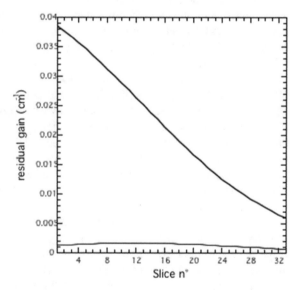

gain between the two passes is only related to the fact that the first slice at the second
pass is the last slice of the first pass and the "initial" gain before the second pass is
the residual gain.

At that point, it is necessary to look back at what has been done in the past and how
and why the laser technology has evolved. The very first laser was made by Maiman
in 1960 [13] and the laser medium has the shape of the rod. Although Frantz and
Nodvik's set of equations came later in 1963 [8], all lasers where based on that early

design i.e. the longer the rod, the higher the output energy. This was true not only for solid-state lasers but although for gas lasers (chemical lasers, excimer lasers, ion lasers and metal-vapour lasers). The rod design has been extensively used until the mid 80's where solid-state lasers were up to 50-cm long, gas lasers were up to 2-m long sometimes even longer. Essentially because of the classical thermal focal length introduce by the rod under pumping conditions but also because of filamentation effects due to high laser intensities, the laser technology moved from the rod to the disk. The first disk amplifier was successfully developed by John Emmett in 1979 [14].

10.5 Achieving Short Pulse Duration

As soon as there is population inversion (10.27), there is gain (10.31) and this gain can be used for amplification. The question is how long the laser system is able to sustain amplification? The answer is as long as the population inversion is possible. This condition will be fulfilled during the time pumping is effective. In real systems, there are losses and losses will counterbalance the amplification process. It can be shown easily with the rate equations that this processes can be taken into account with 3-level and 4-level systems. The optimum pumping time must be of the order or the upper-level lifetime. It is commonly said that the energy is "stored" during this upper-level lifetime.

Managing the losses of the system was the first step to move from a system driven by the self-relaxation process (like the early laser made by Theodore Maiman) to a system controlled by a fast switch (commonly called Q-switch) where the losses of the system are controlled by a fast electronic device. The self-relaxation process will deliver many spikes with microsecond duration and the spike intensity will decrease exponentially. In the Q-switch technique, the losses are controlled until the maximum gain is reached and finally the gain commutation occurs for delivering a single pulse. The shortest Q-switched pulse duration is of the order of a few nanoseconds.

Starting with Frantz and Nodvik's set of (10.38), (10.39), one can include a lot of details about the cavity losses (residual absorption of the gain medium, mirrors reflectivity, Fresnel losses etc.), its length, the gain medium length such that the gain is normalized to the losses. Experimental results and measurements will help to fit at best the pulse duration and the build-up time (of the pulse inside the cavity). Most of the time a parameter run is necessary to define the best quality factor of the cavity depending on the requirements (shortest pulse duration, maximum extracted energy, etc.). Solving this set of equations is straightforward for example with Mathematica [15] (Fig. 10.9).

For achieving even shorter pulse duration (i.e. shorter than a few nanoseconds), the gain medium must have a large linewidth and the conditions according to (10.22) and (10.23) are no longer valid.

Mode-locking was introduced as early as the laser itself in the 60's and pulse duration of tens of picoseconds were obtained. It was shown theoretically that the

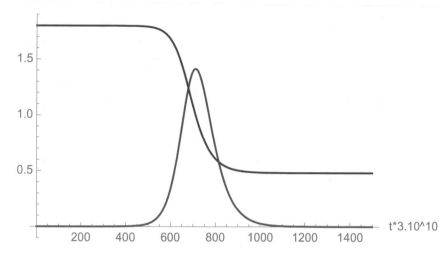

Fig. 10.9 Evolution of the gain and the laser intensity as a function of time during the Q-switch process. Gain (blue curve) is normalized to the gain value at laser threshold which means that when it is equal to 1, laser intensity (red curve, a.u.) is at its maximum. Pulse duration is roughly 7 ns FWHM

minimum duration of mode-locked pulses was of the order of the reciprocal of the oscillation bandwidth of the laser [16]. The basis of the technique is to induce a fixed-phase relationship between the longitudinal modes of the laser's resonant cavity such that constructive interference between these modes will occur at high repetition rate (because the period is equal to the cavity round-trip duration). The laser is then said to be 'phase-locked' or 'mode-locked'. This technique was extensively used in gas lasers, solid-state lasers and dye lasers but a major step was made in 1982 when Peter Moulton discovered the wide bandwidth capability of Titanium doped sapphire [17, 18] covering almost 600 nm (600–1200 nm). With such a large bandwidth, one can create laser pulses at 800 nm as short as a few femtoseconds: less than 5 fs with a spectral Gaussian shape 200 nm full width at half maximum (FWHM).

Unfortunately, amplifying short pulses was not possible because the peak intensity is too high and will cause induced damages to the optical components of the laser system. Another major step was made in 1985 when Donna Strickland and Gérard Mourou [19] were able to amplify a short laser pulse with the chirped pulse amplificationtechnique (CPA). The basic idea is at first to stretch the pulse from few fs to few ns, then amplify it and finally compress it back to almost its original duration. Stretching and compressing the pulse are possible because of the dispersion properties of the classical optical grating: a fist pair of gratings disperses the spectrum and stretches the pulse by a factor of a thousand and a second pair of gratings reverses the dispersion of the first pair and recompresses the pulse. Typical dispersion is around 10 ps/nm such that a 100 nm bandwidth will be stretched to 1 ns. In between those two pairs of gratings, amplification is possible (Fig. 10.10).

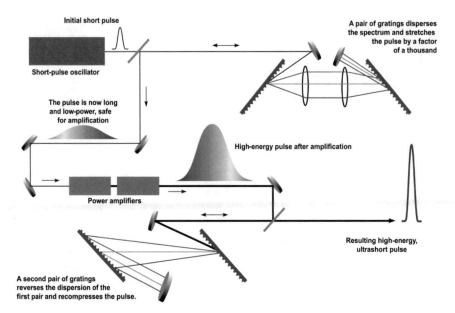

Fig. 10.10 Schematic of beam stretching, amplifying and compressing as used in the CPA technique (adapted from [20])

Introducing the dispersion for modelling the stretched pulse with Frantz and Nodvik's set of (10.38), (10.39) is straightforward. With 20 J and compression to 20 fs, the petawatt is at hand. The CPA technique is used worldwide for building petawatt and multi-petawatt lasers [1, 3] and the actual level is 10 PW [21] while accessing even more energy (10 kJ) with existing fusion lasers will lead to $5 \times 10^{+18}$ W in 20 fs: Zetawatt and Exawatt lasers as discussed by Nobel Prize winner, Mourou [2, 22].

10.6 Conclusion

We have shown that starting from the black-body radiation and the basic relations between the Einstein's coefficients, we are finally able to calculate the fundamental parameter of an amplification system according to the Frantz and Nodvik formalism. With these formulae, it is possible to design many different laser systems from the smallest laser cavity to the biggest multiple pass amplification scheme in fusion lasers. The temporal point of view shall not be forgotten and a split-step method is easy to implement when considering a laser amplifier with a given temporal shape. Chirped pulse amplification is possible too when introducing the spectral-temporal dispersion into Frantz and Nodvik's set of equations.

References

1. C. Danson, D. Hillier, N. Hopps, D. Neely, Petawatt class lasers worldwide. High Power Laser Sci. Eng. **3**(3) (2015). https://doi.org/10.1017/hpl.2014.52
2. T. Tajima, G. Mourou, Zettawatt-exawatt lasers and their applications in ultra strong-field physics. Phys. Rev. Spec. Topics-Accel. Beams **5**, 031301 (2002). https://doi.org/10.1103/physrevstab.5.031301
3. J.H. Sung et al., 4.2 PW 20 fs Ti: sapphire laser at 0.1 Hz. Opt. Lett. **42**(11), 2058–2061 (2017)
4. T. Ditmire et al., CLEO 2014, Technologies for high intensity (STU3F) (2014). https://doi.org/10.1364/cleo_si.2014.stu3f.1
5. M. Planck, *The Theory of Heat Radiation* (P. Blakiston's son & Co. Philadelphia, 1914)
6. A. Einstein, Concerning an heuristic point of view toward the emission and transformation of light. Ann. Phys. **17**, 132–148 (1905)
7. A. Einstein, Zur Quantentheorie der Strahlung. Phys. Z. **18**, 121–128 (1917)
8. L. Frantz, J. Nodvik, Theory of pulse propagation in a laser amplifier. J. Appl. Phys. **34**(8), 2346 (1963)
9. J.B. Trenholme, K.R. Manes, *Simple approach to laser amplifiers, UCRL-51413* (LLNL, Livermore, 1972)
10. R.H. Pantell, H.E. Puthoff, *Fundamentals of Quantum Electronics* (Wiley, New York, 1969)
11. J.E. Murray, W.H. Lowdermilk, The multipass amplifier: theory and numerical analysis. J. Appl. Phys. **51**, 5 (1980)
12. J.E. Murray, W.H. Lowdermilk, Nd: YAG regenerative amplifier. J. Appl. Phys. **51**, 7 (1980)
13. T. Maiman, Stimulated optical radiation in ruby. Nature **187**, 493–494 (1960)
14. Lawrence Livermore National Laboratory, Laser program annual report 1979, UCRL-50021-79 (1980)
15. https://www.wolfram.com/mathematica/
16. P.W. Smith, Mode-locking of lasers. Proc. IEEE **58**(9), 1342–1355 (1970)
17. P.F. Moulton, Ti-doped sapphire tunable solid-state laser. Opt. News **8**, 9 (1982)
18. P.F. Moulton, Spectroscopic and laser characteristics of Ti:Al$_2$O$_3$. J. Opt. Soc. Am. B **3**(1), 125–133 (1986)
19. D. Strickland, G. Mourou, Compression of chirped optical pulses. Opt. Commun. **55**(6), 447–449 (1985)
20. M. Perry, Multilayer dielectric gratings. Sci. Technol. Rev. 25–33 (1995)
21. B. Le Garrec et al., Design update and recent results of the Apollon 10 PW facility, SPIE 10238, 102380Q-1-102380Q-6 (2017)
22. G. Mourou et al., Exawatt-Zetawatt pulse generation and applications. Opt. Commun. **285**, 720–724 (2012)

Chapter 11
Diagnostics of Ultrafast and Ultraintense Laser Pulses

Luca Labate

Abstract In this Chapter, an introductive overview of the experimental methods to characterize an ultrashort laser pulse will be given. Great attention will be paid to the measurement of their time behaviour; this encompasses both the time profile and the pulse contrast. A formal introduction to the mathematical description of an ultrashort pulse in time will be preliminarily provided. Since reaching an ultrashort duration, in the range of 30 fs or less, typically requires an active management of the pulse spectral phase, a particular care will be devoted to the discussion of diagnostics providing both the pulse amplitude and phase. Finally, a brief account will be made of diagnostic techniques to characterize the beam wavefront.

11.1 Introduction

As discussed in other Chapters, the efficient driving of laser-based electron accelerators via the Laser WakeField Acceleration (LWFA) mechanism requires laser pulses with duration in the range of a few tens of femtoseconds. Moreover, although a lot of experiments in such a field have been carried out using longer laser pulses, this requirement also holds for laser-driven protons and light ions acceleration when novel schemes beyond the Target Normal Sheath Acceleration (TNSA) mechanisms have to be exploited (see the Chapter by Borghesi in this Volume). The lowest response time of photodiodes is limited to a few (tens of) picoseconds, both in the leading and trailing edge, basically due to the finite time needed to clean up the depletion zone of the free carriers induced via the photoelectric effect. On the other hand, streak-cameras in the visible and near-IR range enable shorter timescales, of the order of a few hundreds of femtoseconds, to be investigated, which is nevertheless still not enough to directly access the timescale of an ultrashort pulse.

L. Labate (✉)
Istituto Nazionale di Otticam, Consiglio Nazionale delle Ricerche, Pisa, Italy
e-mail: luca.labate@ino.cnr.it
URL: http://www.ino.cnr.it

Sezione di Pisa, Istituto Nazionale di Fisica Nucleare, Pisa, Italy

© Springer Nature Switzerland AG 2019
L. A. Gizzi et al. (eds.), *Laser-Driven Sources of High Energy Particles and Radiation*,
Springer Proceedings in Physics 231, https://doi.org/10.1007/978-3-030-25850-4_11

Therefore, advanced autocorrelation methods have been developed to this purpose, which will be the main topic of this Chapter. As will be shown, unlike "conventional" photodetection methods which typically only provide the (time) intensity profile of the pulse, such advanced techniques may enable the (spectral) phase to be retrieved; as will be quickly mentioned, this can in turn be used to actively compensate for the phase distortions occurring in a Chirped Pulse Amplification (CPA) chain.

Beside the actual main pulse duration, the full knowledge of the time behaviour of an ultrashort laser pulse entails the study of the so-called pulse contrast, that is of the ratio of the main pulse intensity to the one of the laser radiation occurring on much longer timescales before the pulse; this radiation results as a consequence of either the so-called Amplified Spontaneous Emission (ASE), on nanosecond timescales, or of other mechanisms, such as, for instance, an imperfect stretching/compression of the pulse (on tens of picoseconds timescales). This kind of measurement is rather challenging as well, due to the huge dynamic range one has to explore, and requires advanced autocorrelation techniques. A very brief introduction to this subject will be given in this Chapter.

Ultrashort and ultraintense laser systems typically feature a large number of elements, whose optical characteristics and performances can be greatly affected by material non-uniformities, manufacturing imperfections, material stresses, thermally induced deformations and so on. All these factors may introduce wavefront aberrations in the beam, which in turn affect the energy distribution in the focal plane. In order to reach the high intensities (typically in the range 10^{18}–10^{21} W/cm^2) required for laser-driven particle acceleration, the correction of these aberrations needs to be pursued, usually by means of adaptive (deformable) mirrors. Of course, this requires the previous knowledge of the wavefront; methods to perform such a measurement will be briefly dealt with at the end of this Chapter.

Before going on, we observe that any discussion concerning the measurement of "standard" laser pulse parameters, such as, for instance, pulse energy, spectrum and so on, will be left out of this Chapter; rather conventional and well consolidated techniques exist for such purposes, whose discussion can be found elsewhere.

11.2 Experimental Techniques for the Temporal Characterization of Ultrashort Pulses

11.2.1 A Brief Introduction to the Mathematical (and Physical) Description of an Ultrashort Laser Pulse

In this subsection we summarize some mathematical tools and notation which will be useful in the following. We will also quickly review the definitions of standard quantities, in particular in the frequency domain, and discuss their physical meaning. For a deeper introduction to this subject, we refer the reader to [1, 2].

At a fixed point in space, the electric field of a linearly polarized laser pulse can be written as $E(t) = \text{Re}\left(A(t)e^{i\omega_0 t}\right)$, with $A(t) = |A(t)|e^{i\varphi(t)} = |A(t)|e^{i\varphi_0}e^{i\varphi_a(t)}$. Thus, in general $E(t) = \text{Re}\left(|A(t)|e^{i(\omega_0 t + \varphi_0 + \varphi_a(t))}\right)$. The meaning of each term of this expression should be well known:

- ω_0 is the so-called *carrier frequency* (which can be typically identified with the frequency where the spectrum exhibits a maximum).
- $|A(t)|$ is the field envelope, which is proportional to the square root of the instantaneous intensity, $|A(t)| \propto \sqrt{I(t)}$. Most often, depending on the actual situation, the envelope of an ultrashort pulse can be assumed to be of the form $e^{-(t/t_p)^2}$, $\text{sech}(t/t_p)$ or $1/(1 + (t/t_p)^2)$, t_p being related to the pulse duration.
- φ_0 is the so-called *Carrier Envelope Frequency (CEP)*; it can be visualized as the actual phase of the fields with respect to the pulse envelope [ref].
- $\varphi_a(t)$ is an additional (with respect to a linear dependence) time dependent phase function. A non null $\varphi_a(t)$ describes a pulse whose *instantaneous frequency*, defined as $\omega(t) := \frac{d}{dt}\varphi(t) = \omega_0 + \frac{d}{dt}\varphi_a$, varies in time. As we will see in the following, a nonlinear (in t) phase term characterizes what is called a *chirped* pulse.

A complete (and more convenient, as we will see) description of the pulse can be also obtained in the frequency domain. Using Fourier analysis, the electric field and its Fourier transform can be written as

$$E(t) = \frac{1}{2\pi}\int_{-\infty}^{+\infty}\tilde{E}(\omega)e^{i\omega t}\,d\omega \qquad \tilde{E}(\omega) = \int_{-\infty}^{+\infty}E(t)e^{-i\omega t}\,dt \qquad (11.1)$$

As it is well known, under quite general assumptions the knowledge of one of these two descriptions is enough to completely characterize the pulse.

Let us now introduce the so-called *analytic signal*, a quantity widely (and often implicitly) employed for the mathematical description of ultrashort pulses. Since $E(t)$ is real, its Fourier transform is a Hermitian function, $\tilde{E}(-\omega) = \tilde{E}^*(\omega)$, so that the knowledge of the Fourier transform for positive frequencies is sufficient to fully retrieve the signal. We can thus define, for convenience, a new function in the frequency domain, retaining only the positive part of the Fourier transform $\tilde{E}(\omega)$:

$$\tilde{E}^+(\omega) = \begin{cases} \tilde{E}(\omega) & \text{for } \omega \geq 0 \\ 0 & \text{for } \omega < 0 \end{cases} \qquad (11.2)$$

The inverse Fourier transform of this function, $E^+(t) := \mathcal{F}^{-1}[\tilde{E}^+(\omega)]$, is called analytic signal (see [3] for a discussion of its mathematical properties). According to the above observation, $E^+(t)$ is sufficient to retrieve the actual field $E(t)$ (indeed, it is easily seen that $E(t) = 2\text{Re}[E^+(t)]$). From now on, unless otherwise specified, we will use the analytic signal instead of the "real" signal $E(t)$ (sometimes, the $^+$ superscript will be tacitly omitted for convenience). Since $E^+(t)$ is a complex function, it can be written as $E^+(t) = |E^+(t)|e^{i\Phi(t)} = |E^+(t)|e^{i\Phi_0}e^{i\omega_0 t}e^{i\Phi_a(t)} \propto \sqrt{I(t)}e^{i\Phi_0}e^{i\omega_0 t}e^{i\Phi_a(t)}$, where $\Phi_a(t)$ represents any additional phase dependence higher than linear. Of course, the meaning of each term is the same as above. Similarly, in the frequency domain we can write down the Fourier transform of the analytic signal as

$$\tilde{E}^+(\omega) = |\tilde{E}^+(\omega)|e^{-i\phi(\omega)} \propto \sqrt{I(\omega)}e^{-i\phi(\omega)} \tag{11.3}$$

Here, $|\tilde{E}^+(\omega)|$ is the *spectral amplitude* and $\phi(\omega)$ is the *spectral phase*. The spectral amplitude is basically proportional to the square root of $I(\omega)$, the usual spectrum as measured by a spectrometer. The spectral phase is, in essence, the relative phase of each frequency in the waveform. A basic understanding of its meaning in the time domain can be grasped looking at Fig. 11.1 *left*. Here, a superposition of six discrete, equally spaced frequencies ($\omega_k = k\,\delta\omega$) is considered. On the top, the six different waves are supposed to have a zero phase at t_0 ($\phi(\omega_k) = 0$); as a consequence, the overlap of these waves is centered at t_0. On the bottom, the six waves are supposed to exhibit a phase (at t_0) with a linear dependence upon ω: $\phi(\omega_k) \propto k\,\delta\phi$. As it can be seen, the constructive superposition of the waves occurs at a different time; nevertheless, the shape of the pulse remains unchanged. In other words, the spectral phase linear dependence upon ω just results in a pulse delay. However, this doesn't hold anymore for a higher order dependence: in this case, the pulse gets broadened in time, as we will see in a moment.

In general, for a *well defined* pulse the spectral phase can be expanded into a Taylor series around a central frequency ω_0:

$$\phi(\omega) = \sum_{k=0}^{\infty} \frac{\phi^{(k)}(\omega_0)}{k!}(\omega - \omega_0)^k \tag{11.4}$$

with, of course, $\phi^{(k)}(\omega_0) = d^k\phi(\omega)/d\omega^k|_{\omega_0}$. For a more detailed discussion of this expansion we refer to [4]. As it is implicit in our assumptions, each term in the series produces a pulse distortions which is significantly smaller than that of the previous

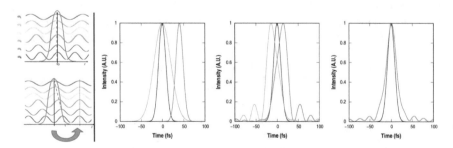

Fig. 11.1 *Left*: Pictorial view of the effect, in the time domain, of the spectral phase exhibiting a linear dependence upon ω. *Right*: Time profiles of pulses with different spectral phase terms. Left plot: original pulse, $\phi^{(1)} = 40$ fs, $\phi^{(2)}/(2!) = 150$ fs^2. Middle plot: original pulse, $\phi^{(3)}/(3!) = +1500$ fs^3 (light green), $\phi^{(3)}/(3!) = -1500$ fs^3. Right plot: original pulse, $\phi^{(4)}/(4!) = 10000$ fs^4

term $(d\phi/d\omega|_{\omega_0}(\omega - \omega_0) \gg (1/2!) d^2\phi/d\omega^2|_{\omega_0}(\omega - \omega_0)^2 \gg (1/3!) d^3\phi/d\omega^3|_{\omega_0}$
$(\omega - \omega_0)^3 \ldots).$

One of the reasons to introduce the concept of spectral phase is easily seen. Indeed, in general, a linear optical system with spectral dispersion acts on the input field by a multiplication by a (complex) transfer function $\tilde{M}(\omega)$ in the frequency domain: $\tilde{E}^+_{out}(\omega) = \tilde{M}(\omega)\tilde{E}^+_{in}(\omega) = \tilde{R}(\omega)e^{-i\phi_{add}(\omega)}\tilde{E}^+_{in}(\omega)$. The amplitude term $\tilde{R}(\omega)$ accounts for a change induced on the pulse spectrum. The spectral phase of the pulse is instead modified according to $\phi_{in}(\omega) \rightarrow \phi_{in}(\omega) + \phi_{add}(\omega)$. Thus, for instance, an initially unchirped pulse, that is a pulse with $\phi''_{in}(\omega_0) = 0$, can acquire a chirp if $\phi''_{add}(\omega_0) \neq 0$.

The three plots of Fig. 11.1 *right* show the effect, in the time domain, of different additional spectral phase terms added to an initially unchirped pulse with a 20 fs FWHM duration. As seen above, a linear phase term $(\phi^{(1)}(\omega_0) \neq 0)$ just results in a time shift of the pulse (*left* plot). As an exercise, let us prove this finding from a formal point of view. We start with an initial unchirped pulse, with a frequency domain representation $\tilde{E}^+_{in}(\omega) = |\tilde{E}^+_{in}(\omega)|$. Adding a linear phase term results in this expression being modified to $\tilde{E}^+_{fin}(\omega) = |\tilde{E}^+_{fin}(\omega)|e^{-i\phi^{(1)}(\omega_0)(\omega-\omega_0)}$. On calculating the inverse Fourier transform, one easily gets

$$E^+(t) = \frac{1}{2\pi}\int_{-\infty}^{+\infty} |\tilde{E}^+_{fin}(\omega)|e^{-i\phi^{(1)}(\omega_0)(\omega-\omega_0)}e^{i\omega t}\, d\omega$$
$$= e^{i\phi^{(1)}(\omega_0)\omega_0} E^+(t - \phi'(\omega_0))$$

This corresponds to a time shift of the pulse of an amount $\phi'(\omega_0)$. The effects of higher order spectral phase terms can be studied in a similar fashion, although resulting in more complicated expressions. Some selected results for spectral phase terms up to the fourth order are shown in Fig. 11.1 *right*. Spectral phase terms of order 2 are usually said to give rise to *Group Delay Dispersion* (GDD), and terms of order 3 to *Third Order Dispersion* (TOD). It is worth noting that 3rd order terms result in additional pulses appearing before or after the initial pulse, depending on the sign of $\phi^{(3)}(\omega_0)$ (see the Figure).

According to the above discussion, we can conclude that for a pulse with a given spectrum (or bandwidth), the shortest duration is reached when no chirp occurs; in the frequency domain, this translates into the spectral phase exhibiting a constant or linear dependence upon ω. In order to provide a simple formal foundation for this claim, we start calculating the pulse duration for a general pulse as

$$\Delta t^2 = \int_{-\infty}^{+\infty} (t - \langle t \rangle)^2 I(t)\, dt = \int_{-\infty}^{+\infty} |(t - \langle t \rangle)E(t)|^2\, dt \qquad (11.5)$$

Using the Plancherel's identity and the equation $\mathcal{F}[(t - \langle t \rangle)E(t)] = ie^{-i\omega\langle t \rangle} \times \frac{\partial}{\partial\omega}(e^{i\omega\langle t \rangle}\tilde{E}(\omega))$, we get

$$= \int_{-\infty}^{+\infty} |\mathcal{F}[(t - \langle t \rangle) E(t)]|^2 \, d\omega = \int_{-\infty}^{+\infty} \left| \frac{\partial}{\partial \omega} (e^{i\omega \langle t \rangle} \tilde{E}(\omega)) \right|^2 \, d\omega$$

and finally, on introducing the spectral amplitude and the phase and calculating the derivative

$$\Delta t^2 = \int_{-\infty}^{+\infty} \left| \frac{\partial}{\partial \omega} |\tilde{E}(\omega)| \right|^2 \, d\omega + \int_{-\infty}^{+\infty} |\tilde{E}(\omega)|^2 \left| \frac{\partial}{\partial \omega} (\omega \langle t \rangle - \phi(\omega)) \right|^2 \, d\omega$$

The first integral is ever positive and depends upon the spectrum. As for the second one

$$\frac{\partial}{\partial \omega} (\omega \langle t \rangle - \phi(\omega)) = \langle t \rangle - \phi'(\omega_0) - \frac{\partial}{\partial \omega} (\text{spectral phase terms } O((\omega - \omega_o)^2))$$

We saw above that the second term accounts, in the time domain, for a pulse delay, so that the first two terms cancels out. Thus, a further (positive) contribution to the time duration exists if the spectral phase exhibits higher order terms (GVD, TOD, …). Ideally, in order to keep a pulse as short as possible, one thus look for optical elements which do not transfer quadratic phase to the pulse or (most of the times) for devices for adjusting/compensating for the accumulated spectral phase.

We conclude this introductory Section mentioning a few basic facts about the spectral phase modifications generally occurring when dealing with ultrashort laser systems.

1. Ordinary transparent media such as those used in laser optics (BK7, fused silica, etc.) feature, under normal dispersion, a $\phi^{(2)}(\omega_0) > 0$, resulting in a positive chirp (lower wavelengths travelling before higher ones). For further details see [5].
2. In their most general configuration, stretcher and compressor induce a positive and a negative chirp, respectively. Moreover, an array of spatial light modulators in the Fourier plane of a compressor can be used to modify the spectral phase of ultrashort laser pulses, thus allowing the pulse shape to be modified in the time domain (see [6]).
3. The accumulated (high order) spectral phase of ultrashort pulses in Chirped Pulse Amplification chains can be compensated for using so-called Acousto-Optic Programmable Dispersive Filters [7], acting on the chirped (stretched) pulse. This is a mandatory technique when intense laser pulses in the range $\lesssim 30$ fs have to be produced.

11.2.2 Basics on 1st and 2nd Order Autocorrelators

In this Section a basic tutorial on the different methods for measuring the duration of ultrashort laser pulses will be given. As it will be clear, this encompasses either the knowledge of just the intensity profile in the time domain, or, using more advanced techniques, of both the spectral amplitude and phase (in the frequency domain), thus accessing the knowledge of the possible pulse chirp. Before entering the discussion, it is useful to highlight the following three basic facts.

- Many of the methods which will be presented in the following requires the measurement of some kind of correlation between two pulses at a variable delay to each other, in a time range of the order of femtoseconds. Such a delay is easily obtained by letting the pulse to be delayed travel longer paths; femtosecond delays require micron-scale optical path lengths, which can be safely and reliably produced and measured using present day technology translation stages and optical encoders.
- Given two pulses with fields $E_{\text{ref}}(t)$ and $E(t)$, it is easy to realize that the experimental measurement of their first order correlation function, defined as

$$C_1(\tau) := \int_{-\infty}^{+\infty} E_{\text{ref}}(t)\, E(t - \tau)\, dt\,,$$

allows $E(t)$ to be retrieved, provided that $E_{\text{ref}}(t)$ (a "reference" pulse) is fully known. It is also easy to understand that, in order for the measurement to provide an accurate estimate of $E(t)$, the length of $E_{ref}(t)$ should be comparable to that of $E(t)$.
- As we mentioned in the Introduction, photodiodes response is limited to the few picoseconds range. When using such a kind of detector in advanced temporal characterization techniques, it is then clear that the measured signal is the integral over a time interval much longer than the actual pulse duration, so that it is proportional to the pulse energy:

$$\text{Read signal} \propto \int_{-\infty}^{+\infty} I(t)\, dt \propto \int_{-\infty}^{+\infty} |E(t)|^2\, dt \propto \text{pulse energy}$$

All these facts will be implicitly used in what follows.

Although not very useful in practice, due to the lack, in general, of a completely characterized ultrashort pulse to be used as a reference, it is worth, from a didactic point of view, to start with simple interferometric techniques. Figure 11.2 shows a conceptual view of the experimental setup to measure the pulse duration of the pulse $E(t)$ using time-domain (*left*) or frequency-domain (*right*) interferometry. In both cases, an ultrashort pulse $E_{\text{ref}}(t)$ is supposed to be available and completely known (that is, in terms of both spectral amplitude and phase).

In the case of time-domain interferometry, a scan of a sufficiently large delay (greater than the pulse duration) is carried out, and the signal corresponding to each delay is recorded. According to what said above, it is easy to verify that the recorded signal is

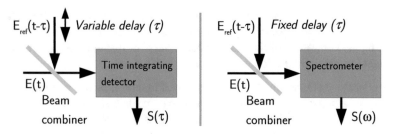

Fig. 11.2 Conceptual optical schemes to measure a pulse duration using time-domain (*left*) or frequency-domain (*right*) interferometry

$$S(\tau) \propto \int_{-\infty}^{+\infty} |E_{\text{ref}}(t - \tau) + E(t)|^2 \, dt = \int_{-\infty}^{+\infty} |E_{\text{ref}}(t - \tau)|^2 \, dt$$
$$+ \int_{-\infty}^{+\infty} |E(t)|^2 \, dt + \left(\int_{-\infty}^{+\infty} E(t) E_{\text{ref}}^*(t - \tau) \, dt + \text{c.c.} \right) \quad (11.6)$$

Notice that the last two terms correspond to 1st order correlation functions. Taking the Fourier transform, one gets $\mathcal{F}(S)(\omega) = A\delta(\omega) + (\tilde{E}(\omega)\tilde{E}_{\text{ref}}^*(\omega) + \text{c.c.})$ (A is a constant), from which the spectral phase of the pulse to be measured can be retrieved (recall that the spectral amplitude is simply related to the pulse spectrum).

In the case of frequency-domain interferometry, the delay is kept at a fixed value, and the spectrum of the overlapping pulses is measured:

$$S(\omega) \propto |\mathcal{F}(E_{\text{ref}}(t - \tau) + E(t))|^2 = \left| \tilde{E}_{\text{ref}}(\omega)e^{i\omega\tau} + \tilde{E}(\omega) \right|^2$$
$$= \left| \tilde{E}_{\text{ref}}(\omega) \right|^2 + \left| \tilde{E}(\omega) \right|^2 + 2\left| \tilde{E}_{\text{ref}}(\omega) \right| \left| \tilde{E}^*(\omega) \right| \cos\left[\omega\tau - \phi_{\text{ref}}(\omega) - \phi(\omega) \right] \quad (11.7)$$

This shows that interference fringes appear in the spectrum with an average fringe spacing inversely proportional to the time delay (which is kept fixed). Again, the spectral phase can be retrieved: indeed, the phase of the fringe pattern yields the spectral phase difference between the reference and the unknown pulse.

Let's come now to the more realistic case in which a second completely characterized ultrashort pulse is not available. In such a case (by far the most common) one has to rely on so-called autocorrelation techniques. The most direct way of getting an autocorrelation trace lies in using a Michelson-like interferometer, to let the pulse to be measured interfere with a delayed replica and directly observing the resulting pulse at different delays (1st order autocorrelation). The conceptual scheme is the same as the one in Fig. 11.2 left, where now $E_{\text{ref}}(t) = E(t)$. Recalling that the field can be written as

$$E(t) = A(t)e^{i\Phi_0}e^{i\omega_0 t}e^{i\Phi_a(t)}, \quad (11.8)$$

the resulting signal as a function of the time delay can be easily seen to be

$$S(\tau) \propto \int_{-\infty}^{+\infty} |E(t-\tau) + E(t)|^2 \, dt \propto 2 \int_{-\infty}^{+\infty} |E(t)|^2 \, dt$$

$$+ 2 \int_{-\infty}^{+\infty} A(t) A(t-\tau) \cos\left[\omega_0 t + \Phi_a(t) - \Phi_a(t-\tau)\right] dt \propto 1 + G_1(\tau) \quad (11.9)$$

where $G_1(\tau)$ is the autocorrelation function. Looking at this expression, we can first notice that the detected signal (trace) is ever symmetric with respect to the time delay τ, even in the case of a chirped pulse. This leads to the conclusion that the spectral phase cannot be measured using this technique. As a matter of fact, the Wiener-Khintchine theorem states that the Fourier transform of the first order autocorrelation function is basically proportional to the pulse spectrum: $\mathcal{F}[G_1(\tau)] = |\tilde{E}(\omega)|^2$ (see, for instance, [8]). Thus, taking the Fourier transform of the signal (with respect to the delay τ) one ultimately only gets the power spectrum. In fact, it can be shown that the full knowledge of $E(t)$ requires the measurement of all the successive $G_n(t)$ [9]. Basically, in a first order autocorrelator the observed trace $S(\tau)$ is only related to the coherence length of the pulse; it is then intuitive that a chirped pulse can result in an apparently shorter (with respect to τ) signal with respect to an unchirped one. Nevertheless, the original pulse duration can be retrieved with some assumptions on the pulse shape and phase; in other words, if one can safely assume that no or negligible chirp is occurring and make a sufficiently accurate guess for the pulse shape (envelope), a first order autocorrelation measurement can provide the pulse duration. We stress, however, that the assumption on the absence of any chirp is rather strong under routine operations of femtosecond lasers.

A certain degree of knowledge of the spectral phase can be gathered using second order autocorrelation techniques, that is looking at the 2nd harmonic signal generated in a nonlinear crystal by two delayed replicas of the ultrashort pulse to be characterized. The most general conceptual scheme of such a kind of autocorrelator is shown in Fig. 11.3 *top*. Since the polarization in the crystal can be written as $P(t, \tau) \propto (E(t) + E(t-\tau))(E(t) + E(t-\tau))$ (see, for instance, [10]), the observed signal on the detector is

$$S(\tau) \propto \int_{-\infty}^{+\infty} \left|(E(t) + E(t-\tau))^2\right|^2 dt$$

$$= \int_{-\infty}^{+\infty} \left|\left(A(t)e^{i\omega_0 t}e^{i\Phi_a(t)} + A(t-\tau)e^{i\omega_0(t-\tau)}e^{i\Phi_a(t-\tau)}\right)^2\right|^2 dt$$

On expanding the square modulus in the integral, one finally gets

$$S(\tau) \propto I_{background} + I_{IA}(\tau) + I_{\omega_0}(\tau) + I_{2\omega_0}(\tau) \quad (11.10)$$

with

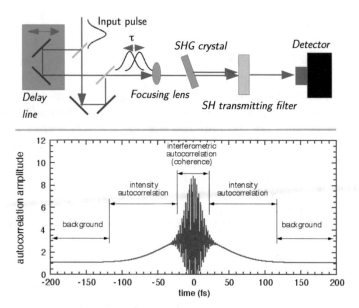

Fig. 11.3 *Top*. Conceptual optical layout of a 2nd order autocorrelator in a collinear geometry. *Bottom*. Typical trace of a 2nd order interferometric autocorrelator (collinear geometry)

$$I_{\text{background}} = \int_{-\infty}^{+\infty} A^4(t)\, dt + \int_{-\infty}^{+\infty} A^4(t-\tau)\, dt = 2 \int_{-\infty}^{+\infty} I^2(t)\, dt$$

$$I_{IA}(\tau) = 4 \int_{-\infty}^{+\infty} A^2(t) A^2(t-\tau)\, dt = 4 \int_{-\infty}^{+\infty} I(t) I(t-\tau)\, dt$$

$$I_{\omega_0}(\tau) = 4 \int_{-\infty}^{+\infty} A(t) A(t-\tau) (A^2(t) + A^2(t-\tau))$$
$$\times \cos\left[\omega_0\tau + \Phi_a(t) - \Phi_a(t-\tau)\right] dt$$

$$I_{2\omega_0}(\tau) = 2 \int_{-\infty}^{+\infty} A^2(t) A^2(t-\tau) \cos\left[2(\omega_0\tau + \Phi_a(t) - \Phi_a(t-\tau))\right] dt$$

The contribution of each term to the observed signal (as a function of the delay τ) can be readily understood looking at the above expressions. In particular, the first term gives a constant background, dependent only on the intensity of the pulse. The second term is due to the SHG due to one photon from the pulse $E(t)$ and one photon from the delayed pulse $E(t-\tau)$. The 3rd and 4th terms give oscillating contributions (as a function of τ) at frequencies ω_0 and $2\omega_0$, respectively. These oscillations are modified by terms involving the difference of the phases of the two replicas at each τ. As an example of a typical 2nd order interferometric autocorrelation trace is shown in Fig. 11.3 *bottom*. As it can be seen, the ratio of the maximum signal, occurring at $\tau = 0$, to the signal at $\tau = \pm\infty$, is fixed; this is a peculiarity of such a kind of autocorrelation, as it can be easily verified from the above equations taking the limits for $\tau \to 0$ and $\tau \to \pm\infty$. The important issue to observe here is that the shape of the fringe pattern strongly depends on the pulse chirp. Indeed, for a given pulse duration, the τ range over which the fringes are visible is related to the pulse chirp: the lower the chirp, the larger is this τ range. On the other hand, the width of the contribution from the 2nd term doesn't depend on the pulse chirp (provided, of course, that the SFG is within the phase matching region of the crystal over the whole delay range). From these reasonings, we can thus conclude that a 2nd order interferometric autocorrelation trace can provide a visual information about the pulse chirp; for instance, in contrast to the one depicted in the Figure, an unchirped pulse would result in a trace where the interval where fringes appears would be as large as the one where the (continuous) signal from the 2nd term (intensity autocorrelation) is visible.

According to these reasonings, one would be tempted to conclude that a 2nd order interferometric autocorrelation could provide a direct measurement of the spectral phase. On the other hand, it is worth observing here that the original pulse giving rise to a given interferometric autocorrelation trace is not unique. Nevertheless, a limited degree of knowledge of the spectral phase can still be retrieved using iterative algorithms (ref. c).

Before concluding this subsection, let us briefly discuss how a 2nd order autocorrelation method can be used to design and implement single-shot devices to measure, under some assumptions, the duration of ultrashort pulses. This is a crucial task as long as low repetition rate laser systems have to be temporally characterized or the pulse duration of an ultrashort pulse have to be minimized using a feedback loop acting on an active spectral phase adjustment devices (see ...). Let's recall that the 2nd term in (11.10) provides a smooth (that is, not modulated) signal as a function of τ, which is related, under some very basic constraints, to the pulse duration only. It is clear that such a smooth signal could be obtained by a full 2nd order interferometric autocorrelation trace by applying a spectral filter to drop out any contribution oscillating at ω_0 and $2\omega_0$. On the other hand, this same effect can be also obtained by slightly modifying the setup shown in Fig. 11.3 *left*, using a different nonlinear generation configuration. In particular, if a scheme is used which involves the two pulses impinging onto the nonlinear crystal at a certain angle to each other (see Fig. 11.4), then the SFG signal generated from one photon per pulse can be discriminated by an angular selection (that is, the SHG signal coming from each of the two pulses separately can be dropped out). This amounts to retaining just the second term in (11.10), that is no background and no interference terms are observed; the trace $I_{IA}(\tau)$ basically provides a 2nd order correlation function, whose width can be used to retrieve the pulse duration. It is to be observed that the 2nd order correlation function is an even function of τ, independent of the symmetry of the actual pulse. Therefore, one cannot uniquely recover the pulse intensity profile. Nevertheless, if one knew a priori that the intensity profile were symmetric, one could recover $I(t)$. Indeed $\mathcal{F}[G_2(\tau)] \propto |\tilde{I}(\omega)|^2$, where $\tilde{I}(\omega) := \mathcal{F}[I(t)]$. If $I(t)$ is symmetric, the its Fourier transform is real and can be therefore recovered knowing its module only. Conversely, if $I(t)$ is not symmetric, the phase of its Fourier transform would be needed as well.

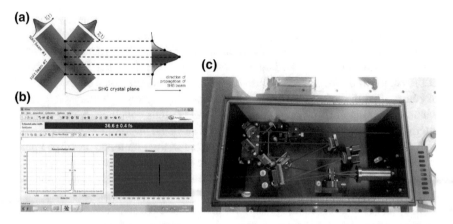

Fig. 11.4 **a** Noncollinear arrangement of the two beams in a nonlinear crystal, resulting in a (time delay)-to-space encoding and thus allowing an implicit scan of a delay range to be obtained with a single-shot. The vertical direction is imaged out onto an area detector, collecting only light at the wavelength of the SFG beam. **b** An example of a raw 2nd order autocorrelation trace (on the right side) and the retrieved intensity profile (left side). **c** Picture of a 2nd order autocorrelator (*Bonzai*, by *Amplitude Technologies*), showing the main optical elements. The beam path is highlighted

A background-free intensity autocorrelator provides a method to measure the pulse duration on a single-shot basis. The underlying concept may be grasped by looking at Fig. 11.4a. The SFG signal coming from a given position of the crystal (along the plane of the two beams) is generated by a (unique) combination of time slices of the two beams; for instance, in the Figure the upper part of the SFG signal comes from nonlinear interaction of the leading edge of the beam 1 and the trailing edge of beam 2. One thus gets an implicit time retardation of the two beams to each other, encoded into the spatial coordinate. By spatially resolving the emission, a single-shot span of the entire delay range of interest is thus obtained, from which a 2nd order intensity autocorrelation trace can be retrieved by taking a simple lineout. An example of a 2nd order single-shot autocorrelator is shown in Fig. 11.4, which also shows the corresponding detected trace and the final retrieved pulse profile.

11.2.3 Advanced Pulse Shape Characterization Techniques in the Time-Frequency Domain

The pulse characterization methods presented so far only enabled a rather limited knowledge of the pulse spectral phase to be obtained; in other words, quite strong assumptions on either the pulse envelope or its chirp had to be made. They were based on measurements carried out either in the time or in the frequency domain. More advanced techniques involve both temporal and spectral resolution at the same time, and are able to provide both the pulse spectrum and spectral phase. As it was mentioned in Sect. 11.2.1, this amounts to know the pulse profile (and thus duration) and its instantaneous frequency (and thus chirp).

One of the mostly used description of an ultrashort pulse in the time-frequency domain involves the use of the so-called spectrogram. Roughly speaking, this basically represents the spectrum of the pulse at each time instant. How can the "instantaneous" spectrum of the pulse $E(t)$ at a given time τ be measured? The general conceptual scheme to accomplish this task is shown in Fig. 11.5 *left*. Starting with the signal $E(t)$, a second "gate" pulse is used to select a very short time slice over which the spectrum can then be calculated (and measured) by the usual Fourier transform. From a formal point of view, we use a gate function $g(t - \tau)$, centered around the time τ, to get a gated signal $E_{gated}(t, \tau) = E(t)g(t - \tau)$; the spectrum of the pulse in a time interval around τ (whose width depends upon the gate function) is then given by the modulus square of the Fourier transform (with respect to t) of this signal:

$$S(\omega, \tau) = \left| \mathcal{F}\left[E_{gated}(t, \tau) \right] \right|^2 = \left| \int_{-\infty}^{+\infty} E(t)g(t - \tau)e^{-i\omega t}\, dt \right|^2 \quad (11.11)$$

A couple of sketches of spectrograms of typical pulses are shown in Fig. 11.5 *right* (see the caption for more details). It is rather intuitive that the knowledge of a spectrogram, that is of the spectrum of the pulse at each time, allows the pulse time profile (including both its amplitude and frequency) to be retrieved. To show it from a formal point of view, let us first observe that from the knowledge of the gated signal $E_{gated}(t, \tau)$ for all the time slices τ the signal $E(t)$ can be retrieved by an integration over τ: $\int E_{gated}(t, \tau)d\tau = \int E(t)g(t - \tau)d\tau \propto E(t)$. Writing $E_{gated}(t, \tau)$ as an inverse Fourier transform (with respect to τ)

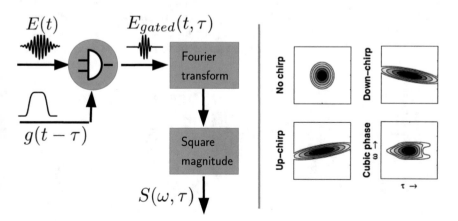

Fig. 11.5 *Left.* Conceptual path leading to measure (and define) the spectrogram of a pulse. *Right.* Sketch of typical spectrograms of ultrashort pulses with characteristic spectral phase terms

$$\tilde{E}_{gated}(t, \Omega) = \int d\tau \, E_{gated}(t, \tau) e^{-i\Omega\tau} \tag{11.12}$$

one easily realizes that the spectrogram is the (square) modulus of the 2D Fourier transform of $\tilde{E}_{gated}(t, \Omega)$:

$$S(\omega, \tau) = \left| \frac{1}{2\pi} \int dt \int d\Omega \, \tilde{E}_{gated}(t, \Omega) e^{i\Omega\tau} e^{-i\omega t} \right|^2 \tag{11.13}$$

The task of reconstructing $E(t)$ is thus moved, via (11.12) and (11.13), to the task of recovering the function $\tilde{E}_{gated}(t, \Omega)$ from its power spectrum $S(\omega, \tau)$ (this is a typical phase retrieval problem). Unlike in the 1D case, in the 2D case this is a (uniquely) accomplishable task (see for instance [11]). In general, iterative algorithms are needed to get $E(t)$ from $S(\omega, \tau)$.

The main experimental challenge in measuring ultrafast optical spectrograms lies in implementing an ultrafast gate. In the *Frequency Resolved Optical Gating* (FROG) method, the pulse is used to gate itself via a nonlinear optical interaction. Different FROG techniques have been developed, relying on different nonlinear effects to accomplish such task; the different FROG implementations thus differ from each other for the gate function, that is the $E_{gated}(t, \tau)$ measured. For a complete description of the different FROG flavours, see [12]. In order to provide a very basic introduction, here we briefly mention two of these variants.

In the Polarization Gating FROG (PG FROG), ultrafast optical gating is achieved through the optical Kerr effect (3rd order nonlinearity). Figure 11.6 *left* shows a conceptual implementation of a PG FROG apparatus. Basically, the "gating pulse" induces a birefringence in the nonlinear crystal and therefore a polarization rotation of the other pulse, which is thus transmitted through the second polarizer. It is easy to realize that in this case the gate is $g(t - \tau) = |E(t - \tau)|^2$, so that the measured (gated) signal is $E_{gated} = E(t)|E(t - \tau)|^2$. It is worth mentioning that PG FROG is one of the most suitable FROG techniques to get single-shot measurement of ultrashort, multi TW laser pulses, using a similar mechanism (time delay encoded in

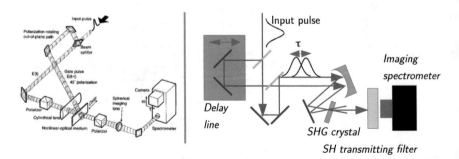

Fig. 11.6 *Left.* Sketch of the optical arrangement of a polarization gating FROG (from [12]). *Right.* Conceptual optical layout of a second harmonic generation FROG device

spatial position) to the one described in the previous subsection. Finally, we want to mention one more FROG scheme, which is more suitable for relatively low-intensity pulses, whose efficiency in giving rise to 3rd order nonlinear processes would be limited. In this case, 2nd harmonic generation ($\chi^{(2)}$ nonlinearity) can be used instead in order to generate the gate pulse. The corresponding device is known as Second Harmonic Generation FROG (SHG FROG). A sketch of the typical optical scheme is shown in Fig. 11.6 *right*. As it can be seen, the conceptual setup is pretty much the same as for noncollinear Intensity Autocorrelation, with spectral resolution added. It is easy to check that the gate pulse is in this case $g(t - \tau) = E(t - \tau)$, so that the measured signal is $E_{gated} = E(t)E(t - \tau)$. Although very simple from the point of view of the experimental arrangement, SHG FROG provides rather unintuitive traces; this is due to the fact that the resulting FROG trace depends directly, roughly speaking, on both the chirp of the pulse to be measured and the chirp of the gate pulse, as it can be realized from the above expression.

We want to conclude this brief discussion observing that FROG is a very powerful and flexible pulse characterization technique; for instance, the same FROG setup may cover a large range of pulse durations. On the other hand, being the reconstruction of the physical quantities from the FROG trace based on iterative algorithms, it is, generally speaking, not very suited for very high repetition rate laser systems.

We want now to briefly tell something about a second method allowing a complete characterization of the time behaviour of ultrashort pulses, namely the *Spectral Phase Interferometry for Direct Electric Field Reconstruction (SPIDER)*. For a complete review of this technique we refer to [13]. The method ca be framed as an evolution of the Interferometric Autocorrelation discussed above, where the two interfering pulses, beside be shifted in time, are also shifted in their central frequency (*spectral shearing*). In order to get an insight of the underlying principle, it is worth, as done above, to look at the measured signal from a formal point of view and then to look at a practical implementation. Starting from the pulse $E_1(t) = A(t)e^{i\omega_0 t}e^{i\Phi_a(t)}$, a replica with both a time and a frequency shift can be written as $E_2(t) = A(t - \tau)e^{i(\omega_0 + \Omega)(t - \tau)}e^{i\Phi_a(t - \tau)}$. The spectrum of the pulse resulting from overlapping these two pulse is then

$$S(\omega) \propto |\mathcal{F}[E_1(t) + E_2(t)]|^2 = |\tilde{A}(\omega)|^2 + |\tilde{A}(\omega + \Omega)|^2$$
$$+ 2|\tilde{A}(\omega)||\tilde{A}(\omega + \Omega)| \cos[\omega\tau + \phi(\omega + \Omega) - \phi(\omega)] \quad (11.14)$$

We observe that the signal shape depends on both the spectral amplitude and phase. Furthermore, for a flat spectral phase, the spectrum exhibits oscillations with a period $2\pi/\tau$. For chirped pulses, the fringe separation depends on the difference between the spectral phases. If the power spectrum is measured with sufficient resolution, the spectral phase can thus be retrieved. Unlike in the FROG case, no iterative algorithm is needed here, the analysis only involving Fourier transforms and signal filtering in the frequency domain.

A conceptual layout of a SPIDER optical setup is shown in Fig. 11.7 *left*. The pulse is first split into two beams. In the upper arm, a replica of the pulse is generated (using,

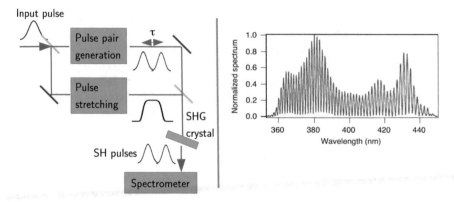

Fig. 11.7 *Left.* Conceptual optical layout of a SPIDER device. *Right.* Typical (raw) spectrum provided by a SPIDER device

for instance, an etalon) with a given (fixed) delay. In the lower arm, the pulse goes through a stretcher, which broadens the pulse in time and gives it a chirp. The pulse pair resulting from the first arm and the stretched pulse then interact in a nonlinear crystal. The crystal phase matching is chosen so as to only allow the SFG from a photon of each of the two replicas and a photon of the stretched pulse. Since the frequency of the stretched pulse at the time of interaction with each of the two replicas is different, this results, as a matter of fact, in a spectral shear of the upconverted pulses with respect to each other; ultimately, thus, two pulses corresponding to $E_1(t)$ and $E_2(t)$ given above are produced. Finally, the power spectrum is measured using a spectrometer. A typical spectrum is shown in Fig. 11.7 *right*. From such a curve, both the spectral amplitude and the spectral phase can be simultaneously retrieved using only direct (non-iterative) algorithms, which makes the SPIDER technique suitable for a time characterization of very high rep rate laser systems.

Before concluding this Section, it is worth to remark that other methods for the complete characterization of ultrashort pulses have been recently developed; among these, we mention GRENOUILLE [14] and WIZZLER [15]. We cannot discuss these methods here, but their physical bases will be easy to be grasped once the above discussion has been followed.

11.2.4 Ultrashort Pulse Contrast Measurement Methods

The very high intensity on target, up to $\sim 10^{22}$ W/cm^2, reached with current CPA laser systems makes the knowledge of the laser intensity profile over an extended time range a mandatory task for the characterization of ultrashort and ultraintense systems. Indeed, since the ionization threshold for most materials lies in the range 10^{10}–10^{12} W/cm^2, even laser emission at an intensity up to ten orders of magnitude

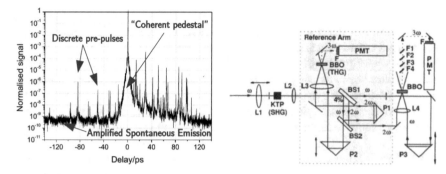

Fig. 11.8 *Left*. Typical intensity profile of an ultrashort, sub-PW class laser pulse over an extended time region before and after the main pulse. *Right*. Original optical layout used to measure the contrast of an ultrashort pulse using 3rd order autocorrelation (from [16])

smaller than that of the main pulse may lead to a target ionization; when this emission occurs before the main pulse, the creation of a pre-plasma may occur and (typically) unwanted physical processes brought forth. On the other hand, CPA laser systems feature a rather characteristic time profile extending over a nanosecond time range.

An example of a typical intensity profile is shown in Fig. 11.8 *left*. The profile includes a ns-long pedestal, due to the Amplified Spontaneous Emission (ASE); pre- and post-pulses are also visible, on timescales of a few up to a few tens of ps, most of the times due to spurious reflections in the amplification chain; finally, a pulse distortion on a timescale of few ps is clearly visible, typically due to an imperfect pulse stretching/compression, leading to spectral amplitude noise or residual spectral phase (see for instance and References therein). The level of these pulse features with respect to the main pulse is broadly referred to as the "pulse contrast". Looking at the figure, it is easy to realize that the knowledge of the laser pulse over an extended time of a few ns is typically required, as well as a very high dynamic range, tentatively spanning 10–12 orders of magnitude or more. With this respect, we first observe that the techniques discussed in the previous subsection are limited to a timespan of a few tens of ps and to a dynamic range of the order of 10^6. An ad hoc approach is thus required to undertake the measurement of the pulse contrast. Usually, high dynamic contrast measurements are performed using high-order autocorrelation techniques. Indeed, the 2nd order autocorrelation discussed above suffers from two main drawbacks when the pulse contrast has to be measured: (a) scattered light from the SHG crystal typically limits the dynamic range to about 10^5; (b) as we observed, the 2nd order trace is always symmetric, so that pulse features before the main pulse cannot in principle be distinguished from those occurring after the main pulse.

The first 3rd order autocorrelation was reportedly performed on the VULCAN laser (at 1053 nm fundamental) at Rutherford Appleton Laboratory [16]. Figure 11.8 *right* shows the optical scheme used in that case. Basically, the pulse is first split into two pulses, one of which is frequency doubled, and delayed to each other; the two pulses are then focused onto a nonlinear crystal whose THG is finally measured, by means of a photomultiplier tube, at different delays. Since, as we saw, the expected signal varies over many orders of magnitude over the timespan to be investigated, different attenuation filters are to be used for different delays in order to get a signal in the useful working range of the photomultiplier tube. Since then, careful optimization methods have been gradually designed and implemented on such a basic scheme in order to enhance the dynamic range, up to the current $>10^{10}$ level; among these, a THG wavelength selection by means of gratings to reduce unwanted radiation, the usage of high dynamic range photomultipliers with variable gain, and the angle optimization (see [17] and References therein).

A 3rd order autocorrelator requires the time interval over which the pulse contrast has to be measured to be scanned by varying the delay between the 2nd and 3rd harmonic beams; it is thus a multi-shot technique, only suitable, in general, for average to high repetition rate laser systems. Efforts have been undertaken, over the past decade, to allow high-dynamics contrast measurements to be carried out on a single-shot base. As we saw in the previous subsection, one of the most direct ways to simultaneously observe an autocorrelation trace at multiple times relies in exploiting the intrinsic time-to-space encoding in a noncollinear nonlinear interaction geometry; a single-shot 3rd order autocorrelator based on this method was first reported in [19]. A different approach was undertaken in [20]. In this case, an optical setup designed on purpose allowed a large number of pulse replicas to be generated, delayed to each other by a few ps. Each of these pulses was then used to interact in a nonlinear crystal at a slightly different angle with the original pulse; the resulting SFG pulses were imaged out onto an area detector. In this way, an interval of $\sim$200 ps was discretely sampled with a single laser shot, with a dynamic range higher than 60 dB. More recently, optical parametric amplification was studied as a tool to get a single-shot autocorrelator to measure the pulse contrast; it is based upon a cross-correlation between a pump and an idler obtained from the pulse to be characterized [21]. The high gain configuration employed allowed the time profile of a low-energy ($<$0.5 mJ) pulse to be characterized within a 50 ps time window with a $>10^7$ dynamic range.

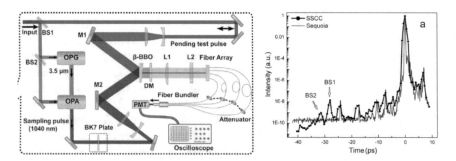

Fig. 11.9 *Left.* Optical layout of the single-shot autocorrelator to measure the pulse contrast as reported in [18]. *Right.* Pulse contrast of a 100 TW class laser system as measured using a *Sequoia* 3rd order autocorrelator by *Amplitude Technologies* (on several shot) and the single-shot autocorrelator shown on the left

Figure 11.9 *left* shows the optical scheme of a single-shot autocorrelator discussed in [18]. In this case, a double stage optical parametric amplifier, pumped by a portion of the pulse to be characterized, was used to produce, via optical parametric generation and subsequent parametric amplification, a very clean pulse; this was then employed to produce a correlation signal via SFG in a BBO crystal. This signal was finally imaged onto a bundle of optical fibers with different length (to sample different time intervals) and different attenuation filters (to account for the very different signal intensities), and finally read using a photomultiplier tube. In this way, a time window of about 50 ps was characterized in a single shot, with a dynamic range in excess of 10^{10}.

11.3   Wavefront Measurement Methods

In this final Section we will briefly deal with the experimental characterization of the (transverse) wavefront of ultraintense laser pulses. As it was mentioned in the Introduction, the beam of an ultrashort and ultraintense laser suffers, in general, from non negligible wavefront aberrations, resulting in a bad quality focal spot unless an active correction by means of deformable mirrors is performed. This is due to the complex optical layout of CPA ultraintense systems, involving a large number of optical elements whose induced wavefront modifications, either due to material imperfections or deformations or to tiny optical misalignments, sum up in the output beam. Thermal gradients induced in the high power amplifiers are also a source of induced wavefront deformations, possibly resulting both in thermal lensing effects and higher-order aberrations. Wavefront distortions can also be introduced by an incorrect alignment of the final focusing Off-Axis Parabola (OAP). As a matter of fact, the focal spot of very short focal length OAPs, typically needed to reach a high intensity in the focus, is very sensitive to very small angle of misalignments. Figure 11.10 *left* shows, for instance, the simulated intensity map for a $f/10$ OAP for very small misalignments, up to 2°. Getting the highest intensity in the focus thus generally demands a wavefront correction compensating the wavefront distortions accumulated in the amplification chain and, possibly, pre-compensating those due to the final optics. Figure 11.10 *right* shows an example of the conceptual layout for such a correction as reported in [22]; the usage of a deformable mirror allowed an ultra-high intensity up to $\sim 10^{22}$ W/cm² to be obtained in the focus with a 45TW laser system. In that case, the wavefront of the focused beam was measured and used in a feedback loop to act on the mirror's actuators. In general, in sub-PW class systems one aims at least at correcting the wavefront aberrations before the Off-Axis Parabola.

Before providing a quick overview of the experimental techniques to characterize the wavefront of a laser beam, it is worth recalling a few formal facts often used for the description of the transverse quality of real beams. Figure 11.11a gives a pictorial view of the effects, in the proximity of the focal plane, of a departure from a perfect (plane) wavefront. Such a departure can be expressed as function

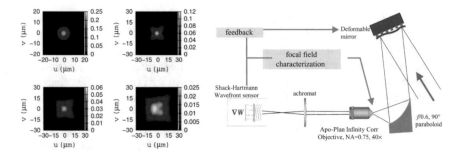

Fig. 11.10 *Left.* Intensity distribution in the focal plane for an $f/10$ OAP, calculated for a perfectly aligned beam (top left) and for increased misalignments: $\vartheta = +1.0°$ (top right), $\vartheta = +1.5°$, $\vartheta = +2.0°$ (bottom right) (from [23]). *Right.* Scheme of the setup used in [22] to correct beam aberrations using wavefront sensing of the focused beam and a deformable mirror

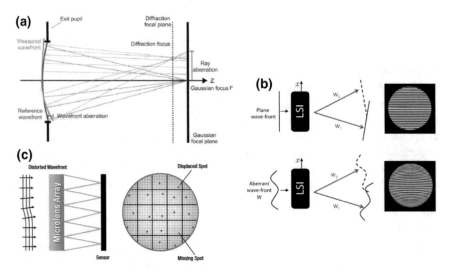

Fig. 11.11 **a** Pictorial ray-tracing of the focusing of a perfect and an aberrated beam. **b** Conceptual scheme of lateral shearing interferometry wavefront sensing. **c** Sketch of the working principle of a Shack-Hartmann wavefront sensor

$\Phi(x, y) = \Phi(\rho \cos \vartheta, \rho \sin \vartheta)$ (we suppose to use a spatial scale such that $0 \leq \rho \leq 1$). It is customary to expand such a function in a series of the so-called Zernike polynomials $Z_n^m(x, y)$. It can be shown (see, for instance, [24] and References therein) that this set of polynomials is the only one with the following properties: (1) is orthogonal over the unit circle; (2) consists entirely of polynomials invariant in form under discrete rotations; (3) contains one polynomial for every integer value of m and n such that $n \geq |m|$ and $n - |m|$ is even. Each Zernike polynomial can be written as $Z_n^m(\rho, \vartheta) = R_n^m(\rho)e^{im\vartheta}$, with

$$R_n^m(\rho) = \frac{1}{2^k k!} \frac{1}{\rho^m} \left(\frac{1}{\rho} \frac{d}{d\rho} \right)^k \left[(\rho^2 - 1)\rho^{n+m} \right]$$

where $k = (n - |m|)/2$. In the aberration theory, only the real part $R_n^m(\rho) \cos(m\vartheta)$ is needed, so that the aberration function can be written as

$$\Phi(\rho, \vartheta) = A_{00} + \frac{q}{\sqrt{2}} \sum_{n=2}^{\infty} A_n^0 R_n^0(\rho) + \sum_{n=1}^{\infty} \sum_{m=1}^{n} A_{nm} R_n^m(\rho) \cos(m\vartheta) \quad (11.15)$$

Here we followed the conventions used in [24]. We warn the reader to check the conventions used, when dealing with equations provided in such a context. For a discussion of the formal properties of the Zernike polynomials we refer to [25]. Moreover, we refer to [8] for a discussion of the kind of aberrations produced by the wavefront distortion represented by each Zernike polynomial.

It is customary to summarize the quality of the wavefront in a single figure, dubbed *Strehl ratio*. It basically gives the maximum intensity achievable with an aberrated beam, normalized to the one from a perfect one (see [8] for a more formal discussion). It can be shown that it is related to the mean square deformation of the wavefront; different approximated formulas are often provided (see for instance [26]). Using one of the expressions provided most often, the Strehl ratio can be expressed as a function of the coefficients A_{nm} of the Zernike expansion as

$$S = \frac{I}{I_0} = 1 - \frac{2\pi^2}{\lambda^2} \sum_{n=1}^{\infty} \sum_{m=0}^{\infty} \frac{A_{nm}^2}{n+1} \quad (11.16)$$

(where, since the indices n, m must satisfy the relationships given above, it is implicit that only the coefficients corresponding to existing Zernike polynomials are to be considered).

Several experimental techniques have been developed for measuring the wavefront of laser beams. Figure 11.11b illustrates conceptually the so-called *Lateral Shearing Interferometry*. This is a self-referencing interferometric method, basically involving the measurement of the interference of 2 (or more, in some variants) replicas of the beam.

For an overview of the measurement techniques used in the field of high power lasers, we refer the reader to [27]. Here we only briefly illustrates the basic principles underlying the so-called *Shack-Hartmann* wavefront sensor, which is likely the most used device. It employs a 2D array of small lenslets (each lenslet with a typical diameter of $\sim 100\,\mu m$) to focus small portions of the beam to be characterized onto an area detector. As it can be realized looking at Fig. 11.11c, the deviation of the position of each spot with respect to the one of a reference beam (that is, a beam with a flat wavefront) is related to the local slope of the deformed wavefront. As a matter of fact, the displacements of the spot centroids with respect to a plane wave reference position is a measure of the local Poynting vector. The experimental data can thus be used to reconstruct the wavefront; in general, the algorithms for analyzing data from a

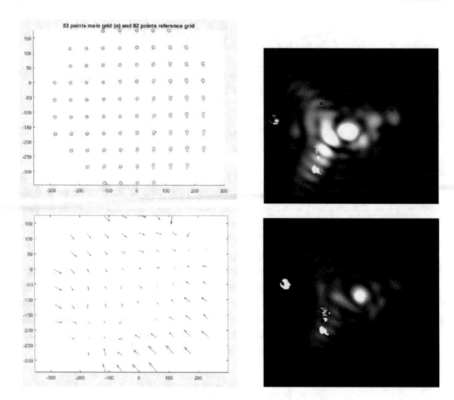

Fig. 11.12 *Left*. Top: positions of the spots from each lenslet of a Shack-Hartmann wavefront sensor (green points); the reference spot positions (that is, the spots of an unaberrated beam) are also shown in red. The displacements between the two set of points are shown on the bottom. *Right*. Measured intensity map in the focal plane of a beam affected by a coma aberration, without (top) and with (bottom) a wavefront correction by a deformable mirror

Shack-Hartmann wavefront sensor provide the coefficients of the Zernike expansion. Furthermore, it can be shown that the M^2 factor (see for instance [28]) of the laser beam can be directly retrieved. For a complete discussion of the data analysis of a Shack-Hartmann wavefront sensor we refer to [29]. As an example, Fig. 11.12 *left* shows the positions of the spots from each lenslet and the displacements from the reference spots in the case of a beam affected by a (negative) defocus (revealed by the radial component of the displacement vectors) and an asthigmatism (revealed by the polar component). As said above, data from a wavefront sensor can be used to actively correct wavefront aberrations using a deformable mirror, possibly with a closed loop algorithm. Figure 11.12 *right* shows, as an example, the intensity pattern in the focal plane for a beam affected (mostly) by a coma wavefront distortion, when focused without and with a wavefront correction by a deformable mirror.

References

1. A.M. Weiner, *Ultrafast Optics* (Wiley, 2009)
2. J.-C. Diels, W. Rudolph, *Ultrashort Laser Pulse Phenomena* (Academic Press, 2006)
3. L. Mandel, E. Wolf, *Optical Coherence and Quantum Optics* (Cambridge University Press, 1995)
4. D.N. Fittinghoff et al., IEEE J. Sel. Top. Quant. Electr. **4**, 430 (1998)
5. I. Walmsley et al., Rev. Sci. Instrum. **72**, 1 (2001)
6. A.M. Weiner, Opt. Commun. **284**, 3669 (2011)
7. F. Verluise et al., Opt. Lett. **25**, 575 (2000)
8. M. Born, E. Wolf, *Principles of Optics*, 7th edn. (Cambridge University Press, 1999)
9. G.B. Giannakis, J. Opt. Soc. Am. A **6**, 682–697 (1989)
10. R.W. Boyd, *Nonlinear Optics*, 2nd edn. (Academic Press, 2003)
11. K. Jaganathan et al., Phase retrieval: an overview of recent developments, in *Optical Compressive Imaging* (CRC Press, 2016). ISBN 9781315371474
12. R. Trebino et al., Rev. Sci. Instr. **68**, 3277 (1997)
13. C. Iaconis et al., IEEE J. Quantum Electr. **35**, 501 (1999)
14. P. O'Shea et al., Opt. Lett. **26**, 932 (2001)
15. T. Oksenhendler et al., Appl. Phys. B **99**, 7 (2010)
16. S. Luan et al., Meas. Sci. Technol. **4**, 1426 (1993)
17. V.A. Schanz et al., Opt. Expr. **25**, 9252 (2017)
18. Y. Wang et al., Sci. Rep. **4**, 1 (2014)
19. J. Collier et al., Laser Part. Beams **19**, 231 (2001)
20. Ch. Dorrer et al., Opt. Expr. **18**, 13534 (2008)
21. R.C. Shah et al., Eur. Phys. J. D **55**, 305 (2009)
22. S.-W. Bahk et al., Appl. Phys. B **80**, 823 (2005)
23. L. Labate et al., Appl. Opt. **55**, 6506 (2016)
24. G.J. Gbur, *Mathematical Methods for Optical Physics and Engineering* (Cambridge University Press, 2011)
25. V. Lakshminarayanan et al., J. Mod. Opt. **58**, 545 (2011)
26. V.N. Nahajan, J. Opt. Soc. Am. **72**, 1258 (1982)
27. H. Wang et al., High Power Laser Sci. Eng. **2**, e25 (2014)
28. O. Svelto, *Principles of Lasers*, 4th edn. (Plenum Press, 1998)
29. B. Schafer et al., Rev. Sci. Instrum. **77**, 053103 (2006)

Index

© Springer Nature Switzerland AG 2019
L. A. Gizzi et al. (eds.), *Laser-Driven Sources of High Energy Particles and Radiation*, Springer Proceedings in Physics 231, https://doi.org/10.1007/978-3-030-25850-4

Printed in the United States
By Bookmasters